# Sustainability of Thermochemical Waste Conversion Technologies

Cynthia Ofori-Boateng

# Sustainability of Thermochemical Waste Conversion Technologies

Springer

Cynthia Ofori-Boateng
Division of Engineering and Computer
Science, Department of Chemical
Engineering
University of Pittsburgh
Johnstown, PA, USA

ISBN 978-3-031-64341-5 ISBN 978-3-031-64342-2 (eBook)
https://doi.org/10.1007/978-3-031-64342-2



This Springer imprint is published by the registered company Springer Nature Switzerland AG
The registered company address is: Gewerbestrasse 11, 6330 Cham, Switzerland

# Preface

Vast varieties of biomass and waste materials, such as municipal wastes and lignocellulosic biomass, are found to be excellent sources of raw materials for cost-efficient biofuels production through thermochemical conversion technologies. Sustainability encompasses the social, economic, environmental, and thermodynamic efficiencies of the production technology. Assessing the sustainability of waste conversion pathways would help pinpoint inefficiencies hence improving the process economically, environmentally, and thermodynamically.

This book discusses the major sustainable potential feedstocks/waste for thermochemical conversion into bioproducts such as biodiesel and bioelectricity. Though there exist many pathways for thermochemical waste conversion (such as combustion, gasification, and pyrolysis) which operate on laboratory, pilot, and commercial scales, their sustainability indices are scarce as there exist few sustainability assessment tools to help pinpoint inefficiencies. This book assesses the sustainability of various types of thermochemical conversion pathways using technoeconomic analysis as well as exergetic life cycle assessment tools. Common sustainability issues and the way forward for sustainable thermochemical waste conversion into bioproducts are detailed in this book. For overall sustainability, thermochemical waste conversion process development alternatives are also discussed in this book.

Sustainability data of biofuel technologies is scarce in literature. Being one of the experts in exergy analysis and sustainability of biofuel processes, addressing technology inefficiencies in a monograph or book using more comprehensive tools like exergetic life cycle analysis would draw the attention of researchers and students to the more important issues surrounding biofuel production sustainability, instead of just developing new biofuels technologies whilst the existing ones are facing challenges.

Given its scope, this is a valuable resource for renewable energy policy makers, bioprocess researchers in academia and related industries, students studying in the fields of Green Chemistry, Chemical and Mechanical Engineering as well as the general publics who have great interest in biofuels for sustainable development. Almost all books on thermochemical biomass conversion address only the process and new technologies, but few tend to address the technical and thermodynamic issues pertaining to sustainability due to the use of fossil fuel in the manufacturing process. This book bridges this knowledge gap, and subsequently outlines specific exergetic improvement options for biofuel and biochemicals production which is scarce in literature. This book assesses the sustainability of bioprocess technologies in a more concise manner for students to understand and apply the knowledge in their future engineering careers.

This book, *Sustainability of Thermochemical Waste Conversion Technologies*, discusses the thermodynamic, environmental, and economic sustainability of biofuels, bioenergy, and biochemicals production from all kinds of waste materials via thermochemical conversion pathways. This book is sectioned into 3 main parts comprising a total of six (6) chapters.

Chapter 1 outlines the detailed overview of the various types of waste materials that are readily available and compatible with thermochemical conversion technologies. Using thermochemical conversion technologies to generate value-added bioproducts from waste materials is a way to manage the large amount of waste causing environmental degradation. The physicochemical characteristics of waste such as Municipal solid waste (MSW), industrial waste, agricultural waste, etc., are discussed, and their chemistry or synergistic effects with thermal and chemical treatments are elaborated.

In every manufacturing plant, there is a need to have a continuous flow of raw materials which would sustain the plant and make it more profitable in the long term. As new innovative biofuel and bioenergy technologies are evolving, feedstock (from first to fourth generations) is constantly being extracted and utilized. Because waste materials are the main feedstock for technologies considered in this book, it becomes imperative to know their availability as they remain the major contributing factors to the successful operation of thermochemical conversion technologies. Chapter 2 reviews the statistics and data of waste generation in the world and their potential to remain available for sustainable production.

Before assessing any technology or process for sustainability, it is important to know the technical process flow regime. Chapter 3 dives into the main conventional thermochemical conversion technologies like pyrolysis, gasification, hydrothermal treatment, etc., and addresses the advantages and disadvantages of each technology.

As research and development advances, there is rapid evolution of innovative new technologies that are improvements to the traditional thermochemical conversion technologies. Chapter 4 elaborates on the current development, drawbacks, and breakthroughs of thermochemical conversion technologies as well as the new advancements in thermochemical waste conversion such as hybrid-renewable energy systems, ultrasonic-assisted technologies, and plasma pyrolysis.

Chapter 5 discusses the literature results of sustainability assessment of biofuels (bio-oil, biodiesel, bio-syngas, biochar, etc.) using exergy analysis, life cycle analysis (LCA), and a combination of both (exergetic life cycle assessment) and gives systematic ideas to help investors and stakeholders to make informed decisions on thermochemical conversion processes.

Lastly, technoeconomic analysis (TEA) of commonly used thermochemical conversion technologies to produce hydrogen, biodiesel, bio-gasoline, and biochemicals are assessed in Chap. 6 based on literature data. Sensitivity and profitability analyses are discussed to know the major factors that inform the cost of products generated via thermochemical waste conversion technologies.

Johnstown, PA, USA Cynthia Ofori-Boateng, Ph.D.

# Acknowledgments

From the depth of my heart, I want to express my sincere gratitude to Elohim, the Almighty God who gave me the wisdom, knowledge, understanding, strength, grace, and mercies through Jesus Christ to start and complete this book. For this book to be published is the doing of the Lord, and it is marvelous in my eyes.

The completion of this book has been a reality because of the immense support from my family. I say a big ‘thank you’ to my lovely children, Caleb and Christabel who always prayed for me and encouraged me to finish well when my strength was almost gone. My sweet mother, Doris, has been a source of inspiration and strength to me, and I say, ‘thank you, mom’. Thanks also go to my dear husband Kenneth for his support.

I thank all my friends and church members for their support and prayers.

Finally, I thank the University of Pittsburgh for providing me with helpful resources to complete this book.

# About This Book

This book elaborates on the sustainability of biofuels and biochemicals production via thermochemical conversion pathways. Sustainability encompasses the social, economic, environmental, political, and thermodynamic efficiencies of a production technology. Assessing the sustainability of waste conversion pathways would help pinpoint inefficiencies hence improving the process economically, environmentally, and thermodynamically. This book discusses the major sustainable potential feedstocks/waste for thermochemical conversion into bioproducts such as biodiesel and bioelectricity. Though there exist many pathways for thermochemical waste conversion (such as combustion, gasification, and pyrolysis) which operate on laboratory, pilot, and commercial scales, their sustainability indices are scarce as there exist few sustainability assessment tools to help pinpoint inefficiencies. This book assesses the sustainability of various types of thermochemical conversion pathways using technoeconomic analysis as well as exergetic life cycle assessment tools. Common sustainability issues and the way forward for sustainable thermochemical waste conversion into bioproducts are detailed in this book. For overall sustainability, thermochemical waste conversion process development alternatives are also discussed in this book.

Given its scope, this is a valuable resource for renewable energy policy makers, bioprocess researchers in academia and related industries, students studying in the fields of Green Chemistry, Chemical and Mechanical Engineering as well as the general publics who have great interest in biofuels for sustainable development. Almost all books on thermochemical biomass conversion address only the process and new technologies, but few tend to address the technical and thermodynamic issues pertaining to sustainability due to the use of fossil fuel in the manufacturing process. This book bridges this knowledge gap, and subsequently outlines specific exergetic improvement options for biofuel and biochemicals production which is scarce in literature. This book assesses the sustainability of bioprocess technologies in a more concise manner for students to understand and apply the knowledge in their future engineering careers.

# Contents

# About the Author

**Dr. Cynthia Ofori-Boateng** is currently an Assistant Professor of Chemical Engineering at the University of Pittsburgh, Johnstown with a BSc., MSc., and Ph.D. in Chemical Engineering. Prior to her current role, she was a Senior Scientist supporting Upstream Process Development Projects at the Center for Breakthrough Medicines at the Discovery Labs and Eurofins Lancaster Laboratories, Inc. with Janssen Pharmaceutical Company of Johnson and Johnson in Pennsylvania, USA.

Cynthia has about 15 years of both academic and industrial experience through which she has been able to achieve excellence through publications, research, scholarships, and teaching. Cynthia was part of a team that designed and installed a 10 $m^3$ biogas plant as well as built a fuel analytical testing laboratory for Koforidua Technical University in Ghana. Whilst at the United States Department of Agriculture (USDA/ARS) in Wyndmoor, Pennsylvania, Cynthia helped her team to develop a new technology (Tail Gas Reactive Pyrolysis/TGRP) which was more sustainable than the existing pyrolysis plant. Cynthia led the team to assess and improve the sustainability of the TGRP plant which could produce biofuel to generate electricity at about 65% cheaper selling price compared to conventional electricity production from crude oil. With the same innovative engineering and entrepreneurial ideas and skills, Cynthia has supervised many groups of students to design pilot and commercial biofuel production plants utilizing waste materials for potential funding, set up, and operation.

Cynthia's Bioproducts Sustainability Research Group also performs laboratory experiments on how to improve the biofuel production plants by optimizing and modeling processes in Aspen Plus software and performing sustainability assessment via the exergetic life cycle (ExLCA) and exergo-economic analyses tools.

Cynthia's main career focus has been on the Sustainability of Biorefineries and Waste Management Technologies. Cynthia has attended numerous international conferences to present her research findings, and has published numerous research articles, a couple of books, and book chapters. Cynthia is a member of the Ghana Institution of Engineers (GhIE), American Institute of Chemical Engineers (AIChE), and American Chemical Society (ACS).

# Abbreviations

| | |
|---|---|
| ABS | Acrylonitrile butadiene styrene |
| AC | Alternating Current |
| ADP | Abiotic Resources Depletion Potential |
| AI | Artificial Intelligence |
| ALCC | Annualized Life Cycle Cost |
| AP | Autothermal Pyrolysis |
| API | American Petroleum Institute |
| BECCS | Bioenergy with Carbon dioxide Capture and Storage |
| BEP | Breakeven Point |
| BPC | By-product Credits |
| BtE | Biomass-to-energy |
| BTEX | Benzene, Toluene, Ethylbenzene, and Xylene |
| BTG | Biomass Technology Group |
| C/N | Carbon to Nitrogen ratio |
| CAD | Canadian Dollar |
| CAGR | Compound Annual Growth Rate |
| $C_{BM}$ | Bare Module Cost |
| CC | Capital Cost |
| CHP | Combined Heat and Power |
| CML | Chain Management by Life Cycle Assessment |
| $CO_2$ | Carbon dioxide |
| CPHEEO | Central Public Health and Environmental Engineering Organization |
| CPW | Compound Present Worth |
| CT | Conventional Torrefaction |
| DC | Direct Current |
| DCCFRR | Discounted Cash Flow Rate of Return |
| DEEP | Department of Energy and Environmental Protection |
| DEFRA | Department for Environment, Food, and Rural Affair |
| DI-CC | Direct Incineration with Carbon Capture |
| DME | Dimethyl Ether |

| | |
|---|---|
| DOE | Department of Energy |
| EC | Energy Cost |
| EPA | Environmental Protection Agency |
| ESP | Electrostatic Precipitators |
| ETSAP | Energy Technology System Analysis Programme |
| EU | European Union |
| FAETP | Freshwater Aquatic Ecotoxicity Potential |
| FC | Feedstock Cost |
| FCC | Fluid Cracking Catalysts |
| FCI | Fixed Capital Investment |
| FTS | Fischer-Tropsch Synthesis |
| GC | Gas Chromatography |
| GCxGC (GC by GC) | On-line comprehensive two-dimensional Gas Chromatography |
| GH | Gas Heater |
| GHG | Greenhouse Gas |
| Gt | Gigaton |
| GWP | Global Warming Potential |
| H-BEA | Homemade beta zeolite |
| HDPE | High Density Polyethylene |
| HHV | High Heating Value |
| HRSG | Heat Recovery Steam Generator |
| HT | Hydrothermal Treatment |
| HTL | Hydrothermal Liquefaction |
| HTP | Human Toxicity Potential |
| HT-PEMFC | High Temperature Proton Exchange Membrane Fuel Cell |
| IBP | Initial Boiling Point |
| IEA | International Energy Agency |
| IP | Intermediate Pressure |
| IPCC | Inter-governmental Panel on Climate change |
| ISO | International Standard Organization |
| LC | Levelized Cost |
| LCA | Life Cycle Analysis |
| LCC | Life Cycle Cost |
| LCI | Life Cycle Inventory |
| LCIA | Life Cycle Impact Assessment |
| LDPE | Low Density Polyethylene |
| LLC | Limited Liability Company |
| MAETP | Marine Aquatic Ecotoxicity Potential |
| MC | Maintenance Cost |
| MLW | Municipal Liquid Waste |
| MSW | Municipal Solid Waste |
| MT | Microwave Torrefaction |
| MW | Molecular Weight |
| NCG | Non-Condensable Gases |

| | |
|---|---|
| NMVOC | Non-Methane Volatile Organic Compounds |
| NPV | Net Present Value |
| NPW | Net Present Worth |
| NRTL | Non-Random Two Liquid |
| OC | Operating Cost |
| ODP | Ozone layer Depletion Potential |
| OECD | Organization for Economic Co-operation and Development |
| PEF | Product Environmental Footprint |
| PET | Polyethylene Terephthalate |
| POCP | Photochemical Ozone Creation Potential |
| PP | Polypropylene |
| PP | Payback Period |
| PS | Polystyrene |
| PSA | Pressure Swing Adsorption |
| PVC | Polyvinyl Chloride |
| PWF | Present Worth Factor |
| R&D | Research and Development |
| RFS | Renewable Fuel Standards |
| SCCP | Solar Char-Cycling Pyrolysis |
| SCWG | Supercritical Water Gasification |
| SDP | Solar Direct Pyrolysis |
| SG-CLC | Steam Gasification-Chemical Loop Combustion |
| SNG | Synthetic Natural Gas |
| SV | Salvage Value |
| $tCO_2$ | Ton Carbon Dioxide |
| TEA | Technoeconomic Analysis |
| TETP | Terrestrial Ecotoxicity Potential |
| THS | Thermochemical Hybrid Waste-to-energy Systems |
| TPC | Total Production Cost |
| TRACI | Tool for the Reduction and Assessment of Chemical and other environmental Impacts |
| UK | United Kingdom |
| US | United States |
| WC | Working Capital |
| WHO | World Health Organization |
| WtE | Waste-to-Energy |

## Unit of Measurement

| | |
|---|---|
| A | Ampere (coulombs/second) |
| Bbl | Billion barrels |
| $CO_{2eq}$/ton | Carbon dioxide equivalent per ton |

| | |
|---|---|
| EJ | Exajoule |
| g/min | Gram per minutes |
| kg | Kilogram |
| kHz | Kilohertz |
| kW | Kilowatt |
| L/min | Liter per minute |
| MJ | Megajoule |
| mPts/kg | Milli-ecopoints per kg |
| Mt | Metric ton |
| MW | Megawatt |
| $MW_e$ | Megawatt electricity |
| ng/g | Nanogram per gram |
| tons/h | Tons per hour |
| TWh | Terawatt-hour |
| $W/m^2$ | Watt per square meter |
| μPts/kg | Micro-ecopoints per kg |

# List of Symbols

| | |
|---|---|
| $Ex_{ch}$ | Chemical Exergy |
| $Ex_{work}$ | Exergy of a system due to work production |
| $Ex_i$ | Exergy of *i*th component |
| $m_i$ | Mass of *i*th component |
| $S_{generation}$ | Entropy Generation |
| $Ex_{destruction}$ | Exergy destruction |
| $Ex_{mass,in}$ | Material Exergy Input |
| $Ex_{mass,out}$ | Material Exergy Output |
| $(c^2/2)$ | Kinetic exergy |
| $(gx)$ | Potential exergy |
| $C_i$ | Initial investment cost |
| $C_s$ | Single future cost |
| d | discounted rate |
| $Ex_{\mathrm{nu}}$ | Nuclear exergy |
| $Ex_{\mathrm{ph}}$ | Physical exergy |
| $I$ | Irreversibility |
| i | Number of years |
| n | Number of years for payment |
| $P_{Vr}$ | Recurring annual payment |
| $P_{Vs}$ | Single future payment |
| r | Interest rate |

# List of Figures

# List of Tables

# Part I
# Feedstocks For Thermochemical Conversion Into Bioproducts

# Chapter 1
# Types of Waste Materials for Thermochemical Conversion into Bioproducts

**Abstract** Due to their potential applicability of supporting waste management worldwide, thermochemical waste conversion technologies such as pyrolysis, combustion, and hydrothermal processing are currently reported as the most promising routes for sustainable green fuels, bioenergy, and biochemicals production. Extensive diversity of carbonaceous waste materials such as municipal solid wastes (MSW), industrial hazardous and non-hazardous wastes, and agricultural residues (e.g., lignocellulosic biomass) are excellent sources of raw materials for cost-efficient production of biochemicals, biofuels, heat, and electricity through thermochemical conversion technologies whereby the wastes are combusted instantaneously at specific process conditions. The types of biomass wastes and other potential waste materials are characterized predominantly by their physicochemical properties which eventually determine the quality, calorific value, and quantity of bioproducts generated via thermochemical conversion pathway. Thus, the conversion of waste materials would be efficient when the suitable type of waste is converted via the appropriate thermochemical conversion technology. This chapter discusses the major potential carbonaceous feedstocks/waste that are outstanding for thermochemical conversion into bioproducts.

**Keywords** Biomass · Waste materials · Thermochemical conversion technology · Sustainability

## 1.1 Thermochemical Conversion as Waste Management Strategy

Since time immemorial, human beings have not ceased to consume natural resources and generate wastes in large quantities without proper management which eventually affects lives physically especially with rapid urbanization, high population growth, and high standards of living. This cycle of 'consume and generate' is not going to end so far as human beings still exist on earth. Industrial technologies and domestic activities have been the major contributors of waste generation in the world with



C. Ofori-Boateng, *Sustainability of Thermochemical Waste Conversion Technologies*,
https://doi.org/10.1007/978-3-031-64342-2_1

about 80–90% share of total global wastes in 2021 (OECD 2021). Waste generation associated with manufacturing can result from technical, technological, and design inefficiencies and thermodynamic/exergy destruction (e.g., inefficient equipment increases entropy generation hence dissipating energy as waste) which directly affect the sustainability of the process and products. Thus, there is an urgent need to find sustainable solutions to the problem of waste generation and management associated with production and consumption of natural resources.

Fossil energy sources, such as crude oil, diesel, and natural gas, cannot be reliable sources to meet the increasing electricity demands for future generations due to their exhaustible nature and carbon emissions. To reach the goal of net zero $CO_2$ emissions in 2050, it becomes imperative to resort to sustainable source of energy (IEA 2021). Thus, the rising energy consumption coupled with high fossil fuel prices globally, are gradually shifting the world into green fuels which are environmentally friendly. Many countries have taken various initiatives to implement stringent emission requirements by the reduction of greenhouse gases (GHGs) into the atmosphere.

The idea of sustainability and sustainable development which defies economic instability, environmental degradation, societal malfunctioning, and technological inefficiencies was dated back to the early 1990s when most countries faced resource management challenges (Beder 1996). Sustainable production involves the efficient utilization of renewable and safe materials to create cost effective products through safe technologies that do not destroy the environment. Biofuels, biochemicals, and bioenergy from biomass waste and other waste materials would be highly sustainable when the production technologies are improved.

Though few biomass-to-energy (BtE) or waste-to-energy (WtE) conversion technologies have been commercialized, sustainable production and consumption of biofuels and bioenergy via these processes have not been fully realized due to gaps or deficiencies in environmental, technical/thermodynamic, economic, and socio-political frameworks (Ofori-Boateng 2022). The release of air pollutants[1] from fossil fuel combustion during manufacturing (due to insufficient renewable or green energy resources), the ineffective management of solid and liquid by-products, the use of toxic chemicals as well as the high cost of production are a few challenges impeding the sustainability of waste-to-energy technologies presently. As a result, there has been extensive research and development (R&D) on both laboratory and pilot scales to improve the sustainability of biomass conversion technologies.

The reduction of waste generation in industries through efficient waste disposal methods whilst the conversion of recyclable wastes into valuable biproducts like bio-fertilizers are some proven sustainable ways to maintain eco-friendly production and consumption (Lee and Ofori-Boateng 2013). The sustainability framework for Biofuels Production outlined by the Lowell Center for Sustainable Production (LCSP 1998) points out some critical principles, criteria, and indicators for sustainable

[1] From the sixth assessment report of the Inter-governmental Panel on Climate change (IPCC), manufacturing plants emit about 20 Gt $CO_2$ annually which forms about 34% of total global carbon dioxide emissions (Dhakal et al. 2022).

agriculture to conserve biodiversity and minimize waste generation. The revised version of Article 4 of the EU Waste Framework Directive (Directive 2008/98/EC) outlines five steps for dealing with wastes under environmental development referred to as the '*waste hierarchy*' namely: prevention, preparation for reuse, recycling, material and energy recovery, wastes disposal (DEFRA 2012). Some countries like the United Kingdom have adopted and have been implementing these criteria for the generation and management of their waste (DEFRA 2012).

The management of waste is crucial in minimizing its detrimental effects on human beings and the environment. One major way to sustainably manage wastes is to transform them into value-added bioproducts such as biofuels whereby their exergy contents[2] are improved (Lee and Ofori-Boateng 2013). The conversion of materials including biomass waste to energy can be achieved through three main pathways namely biochemical (e.g., anaerobic digestion), thermochemical (e.g., pyrolysis), and physiochemical (e.g., transesterification) processes. Out of the various biofuel production technologies, thermochemical conversion processes such as pyrolysis, hydrothermal liquefaction, gasification, etc., have realized immense breakthroughs in R&D due to their conversion efficiencies (Kariim et al. 2022).

Thermochemical biomass waste conversion is an energy production technology which uses heat to convert the organic matter in biomass and other waste materials (such as plastics) into liquid, gaseous, and solid biofuels (such as biodiesel, biogas, and biochar) at high pressures in reactors (Kariim et al. 2022). Other potential energy from this technology includes heat, steam, electricity, and transportation fuels (such as ethanol, diesel, pyrolysis oil, etc.).

Thermochemical waste conversion technologies include pyrolysis, torrefaction, carbonization, gasification, combustion, hydrothermal liquefaction, etc. Though the development of these technologies is ongoing, there are a couple of commercialized biofuel production plants in most parts of the world including the United States, Europe, and Asia (Srivastava et al. 2023). For instance, according to BTG Bioliquid company,[3] in 2015, commercial biomass pyrolysis plants in the Netherlands (Empyro Hengelo), Finland (Green Fuel Nordic), and Sweden (Pyrocell) with production capacities of about 20 million liters of bio-oil annually began operation (IEA 2021). Sustainability assessment of these manufacturing plants would drastically improve the environment, cost, and technical efficiency of process equipment.

Thus, it becomes critical to manage waste efficiently in manufacturing industries, and this can be achieved by waste-to-energy conversion, recycling, and reuse. R&D on waste to energy via the thermochemical pathway has focused on a few technologies and waste material types. However, for a complete assessment of all potential thermochemical conversion technologies and waste material types may pinpoint the more

[2] Exergy content of a resource is the quality of energy within the resource that is available to do work.

[3] BTG stands for Biomass Technology Group in Netherlands which has specialized in thermochemical biomass conversion technologies for almost four decades.

sustainable ways of producing biofuels and biochemicals. This chapter discusses all potential sources of waste materials that are suitable for thermochemical conversion into fuels, energy, and chemicals.

## 1.2 Types of Waste Materials for Thermochemical Conversion into Bioproducts

There are vast varieties of waste materials which can be transformed into value-added bioproducts via thermochemical conversion technologies. Generally, these waste materials can be categorized into biodegradable and non-biodegradable wastes. Both categories of waste can be converted into value-added bioproducts via thermochemical conversion technologies. However, the quality and quantity of product recovered from these wastes can differ based on the characteristics (both physical and chemical) of the waste material. Waste as feedstock for thermochemical conversion into value-added bioproducts can be classified according to their origin, composition, and utilization. The subsequent sections elaborate on the types of waste commonly used with thermochemical conversion technologies.

### *1.2.1 Agricultural Wastes*

Agricultural wastes comprise undesirable plant materials or residues generated from agricultural and forestry processes or activities such as growing, maintaining, and harvesting crops. Agricultural waste, which forms about 80% of cultivated plants is generally considered unsuitable for human and animal consumption unless they have been appropriately processed (OECD-FAO 2015). Agricultural waste is easily reachable, collectable, and usable as a source of sustainable energy and biochemicals. Agricultural wastes are renewable and readily available at little or no cost. Because the combustion and landfilling of agricultural waste releases GHG emissions, there may be significant environmental impacts on soil, water, air, and the ecosystem at large if they are not managed properly. Some kinds of agricultural waste can be utilized directly in farms as mulch, fertilizers, compost, etc., as sustainable means of managing agricultural waste.

Agricultural waste can be ***organic*** or ***inorganic*** depending on the source of generation as well as the physicochemical composition. Organic agricultural wastes are derived from organic farming activities and are readily available for sustainable transformation into energy and chemicals. Examples of organic agricultural waste which can be converted into valuable products via thermochemical conversion include:

- crop residues (e.g., rice straw, corn stover, barley straw, oat straw, etc.); agricultural by-products and industrial processing wastes (e.g., plant stems, sugarcane bagasse, oil palm fronds or branches, palm empty fruit bunches, fallen leaves, buds, flowers, trimmings, etc.) generated from the harvest of maize, grains and cereals, sugarcane, etc.; fruit bearing trees (e.g., palm tree); shrubs; vegetable/fruit crop wastes from harvesting (e.g., apples, cabbage, lettuce, tomatoes, etc.); trimmed grass from lawns and gardens, etc. Crop residues form about 80–90% of agricultural waste materials (Smil 1999).
- animal manure generated from animal or livestock rearing (e.g., cattle dung, swine manure, animal fats, etc.).
- Spent mushroom compost generated during the growing of mushrooms.

Inorganic agricultural wastes are the by-products or residues resulting from agricultural activities which involve chemicals and materials that do not come from organic origin. These waste materials include:

- Irrigation wastewater and runoff from agricultural fields which may contain chemical or inorganic components.
- Excess or unused inorganic fertilizers and herbicides.

Figure 1.1 summarizes the main classes of organic and inorganic agricultural wastes.

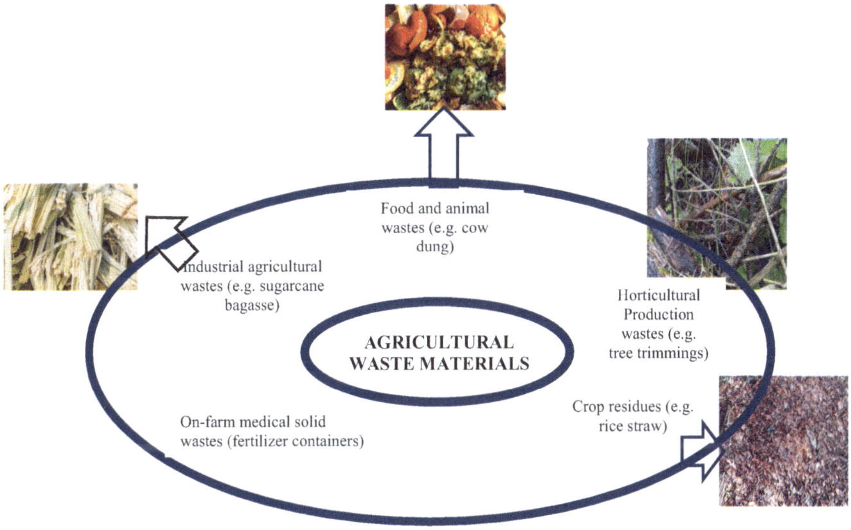


**Fig. 1.1** Pictorial diagram of the classes of agricultural wastes (*Pictures taken by Author*)

## 1.2.2 *Municipal Wastes*

Municipal wastes are complex wastes generated by households, institutions (e.g., schools), companies, and businesses established within a municipal area. Municipal waste composition differs from one location[4] to another and can be classified as solid and liquid wastes based on the composition of the waste.

### 1.2.2.1 Municipal Solid Waste (MSW)

Municipal Solid Waste (MSW) (sometimes called garbage, refuse, or trash) refers to assorted mixture of solid waste materials from households/residential sources, institutions sources, small businesses, corporate offices, and 'occasional' industrial facilities established in a specified municipality. MSW can be grouped according to the following components based on recycling rates and geographical locations:

- Solid Food Waste: This is the largest organic composition of MSW which forms about 20–22% of total MSW generated annually (OECD 2021). Depending on the geographical location of MSW collection, living standards, and way of life of people in a specific municipality, food waste may comprise spoilt food, restaurant and households' leftovers, peels, trimmings of food crops, etc.
- Plastics: These include plastic food packaging (like polyethylene terephthalate (PET) bottles, containers, and jars), disposable utensils, cups and plates, plastic bags, foamed plastics (also called expanded or cellular polymers like sponges), etc. MSW plastics form about 11–16% of total MSW generated per year (Bodzay and Banhegyi 2017; Letshwenyo and Kgetseyamore 2020) comprising about 23% low density polyethylene[5] (LDPE), about 19% high density polyethylene[6] (HDPE), about 9% polystyrene[7] (PS), about 10% polyethylene terephthalate (PET), about 6% polyvinyl chloride[8] (PVC or vinyl), about 14% polypropylene[9] (PP), and about 19% other plastics (e.g., polycarbonate, polyamide, acrylonitrile butadiene styrene (ABS)).
- Papers and Cardboards: These components of MSW are generated in the form of paper products such as cardboard packaging (e.g., cereal boxes, tissue boxes, disposable paper bags, etc.), shipping materials (e.g., corrugated cardboard boxes), paper towels & tissue papers, magazines, newspapers, unwanted mails (e.g., catalogues, flyers), used office papers (e.g., invoices, reports, documents, envelops, shredded papers), etc. MSW papers and cardboards form about 23–29%

---

[4] Based on population size, waste collection pattern, type of people or institution, the composition of municipal waste may vary.

[5] LDPE are used in packing films and bags.

[6] HDPE are used in plastic bottles, containers, and thick bags.

[7] PS are used in expandable disposable food containers.

[8] PVC are used in pipes, flooring, fittings, medical devices, etc.

[9] PP are used in automobile parts, food containers, packaging films, etc.

of total MSW generated annually (EPA 2018) which is the major component of MSW.

- Wood: MSW wood which forms about 6–8% of total MSW generated annually, comprises different forms of wood-based materials such as used or old furniture, construction and demolition woody debris, yard trimmings, crating, pallets, etc.
- Glass: MSW glass refers to the glass component of MSW which is about 4–7% of the total MSW generated per year (EPA 2018). This includes different types of glass products such as jars, containers, bottles, glassware, etc., which are observed as unbroken, broken, or partially broken. Glass forms about 4–5% of the total MSW components.
- Rubber, Textiles, and Leather: The predominant source of rubber in MSW is rubber tires from automobiles which is about 2–3% of total MSW generated annually. Leather waste basically comprises dried and processed animal skin trimmings, scraps, worn out shoes, bags, and jackets, and it forms about 1% of total MSW generated annually. Textile wastes are unwanted, torn, or used textile materials such as clothing, fabrics mostly made of synthetic fibers, etc., which form about 5–6% of total MSW generated annually (EPA 2018).
- Metals: This category of MSW forms about 8–10% of total MSW, and it comprises light and heavy metals (e.g., ferrous metal, copper, brass, bronze, lead, zinc, aluminium, steel, etc.) disposed as wastes from municipalities (Munawar et al. 2023). Metal waste in MSW is mostly found in the form of unwanted appliances, packaging cans, foils, wires, construction materials, plumbing fixtures, musical instruments, electronics, batteries, etc.
- Yard trimmings: These are organic waste materials obtained in the form of leaves, tree branches, grass trimmings, etc., through activities such as landscaping, gardening, yard development, etc., 12–13% of total MSW (Munawar et al. 2023).

MSW does not include construction and demolishing debris which are usually classified as industrial waste (see Sect. 2.3).

#### 1.2.2.2 Municipal Liquid Waste (MLW)

Liquid waste from municipalities is collectively called municipal liquid waste (MLW). MLW is the wastewater or sewage mostly sourced from stormwater runoffs, grease and grit traps, septic tanks and pipes, municipal wastewater treatment plants, residential sewage treatment tanks, sewage, and sewage sludge. MLW usually comprises organic and fecal matter, dissolved and suspended solids, heavy metals, pathogens, and pollutants which can pose environmental and human health risks if not sustainably treated or recycled (Liew et al. 2022). The main components of MLW are sewage and sludge which are high potential sources of feedstock for fuel and energy production via thermochemical conversion technologies. MLW is usually collected via a network of underground pipes (called sewer lines), combined sewers, and separate sanitary sewers which are then transported to treatment plants to remove most contaminants before discharged into the environment. In some cases, sewage sludge is treated in conventional waste disposal facilities like incineration and landfills or used as manure during agricultural activities (Hale et al. 2012). Nonetheless, there are effective ways of sustainably transforming sewage sludge into value-added bioproducts such as thermochemical conversion technologies which are discussed in this book.

Sewage is wastewater generated from municipalities which can be classified as blackwater, or greywater based on their characteristics and origin. Greywater is wastewater originating from household sinks, showers, bathtubs, laundry activities, and kitchen wastewater, and they constitute about 65–100% of total wastewater discharged by households (Morel and Diener 2006). Blackwater originates from toilets, and they contain bulk portions of organic matter and pathogens (Paulo et al. 2013). Though blackwater is generated in lesser quantities (about 30% of MLW) compared to greywater, it is highly hazardous and may pose serious environmental concerns if not properly treated and disposed of (Hernández et al. 2007). The possibilities of generating biofuels and biochemicals from sewage sludge are discussed in detail in Chaps. 3 and 4 of this book.

According to Li et al. (2024) and Wei et al. (2018), municipal sewage sludge can be grouped into primary sludge (PS) and waste activated sludge (WAS) when collected in primary sedimentation tanks and biological treatment tanks, respectively.

### *1.2.3 Industrial Waste*

Industrial waste (also called manufacturing waste) refers to all types of unwanted materials, waste, or residues generated by industries from all their industrial processes or activities such as the metallurgy, construction and demolition, food processing, pharmaceutical/chemical industries, etc. Petroleum distillates, asphalt wastes, and used oils from refineries and construction industries are considered industrial wastes. Industrial waste can be classified as (Sahay 2023):

- Solid (e.g., metal scraps, old medical equipment, construction and demolishing wastes, electronic and electrical equipment waste), liquid (e.g., industrial effluent like food and beverage waste, textile liquid waste, lubricating waste oil, etc.), or gas (e.g., industrial waste gases such as carbon dioxide, oxides of nitrogen, etc., which arise from operation of outdated process technologies, incomplete combustion, etc.)
- hazardous (e.g., industrial effluent, carbon dioxide, chemical waste, sulfur dioxide)
- non-hazardous (e.g., chemical waste)
- Biodegradable which can be broken down into non-hazardous materials with time by means of suitable micro-organisms. Examples of biodegradable industrial waste are animal bones, leather waste, textile waste, herbal leaves, etc.
- non-biodegradables which are the non-degradable industrial wastes hence major contributors of environmental pollution. Non-biodegradable waste and all kinds

of industrial waste may give rise to all sorts of pollution (air, water, land) hence posing serious risks to the ecosystem if not disposed well or recycled. Examples of non-biodegradable industrial waste are plastic waste, radioactive waste, iron, other metallic waste, etc.

The majority of industrial waste originates from the food processing industries, metallurgy, and construction industries, and their composition may vary from industry to industry. Other companies that contribute to industrial waste generation globally include the automotive and transportation industries, government agencies, universities, healthcare and medical offices, retail shops, utilities companies, printing and publishing companies, consulting and engineering units, and chemical manufacturing companies. Based on the physicochemical compositions, most industrial waste materials are considered hazardous according to the guidelines of the Environmental Protection Agency (EPA). Most hazardous industrial waste from chemical and biomedical activities (e.g., adhesives, paint wastes, etc.) are not suitable for thermochemical conversion into chemicals and fuels.

Construction and demolition wastes include toxic, non-toxic, or flammable wreckage, debris, concrete, broken bricks, stones, glasses, and metal scraps or refuse generated from activities such as demolition, land clearing, and wrecking. Construction and demolition waste can also be classified into hazardous, non-hazardous, and specialty wastes, and they form about 17–19% of industrial waste (DEEP 2016). Non-hazardous demolition and construction wastes such as carpentry and floorwork debris, plastics, concrete, and stones can be used as feedstock for thermochemical conversion into valuable products.

## 1.3 Characteristics and Chemistry of Waste Materials for Thermochemical Conversion

For a feedstock to be an excellent source of raw material for thermochemical conversion into energy, fuels, and chemicals, certain important characteristics and compositions must be justified. During thermochemical conversion of waste materials, chemical bonds between adjacent carbon, hydrogen, and oxygen molecules break down to liberate their chemical energy stored in them. The chemistry of waste materials as feedstock for thermochemical conversion is very critical in selecting the right conversion technology and process conditions.

The ultimate[10] or elemental composition (e.g., carbon, hydrogen, nitrogen, oxygen, etc.), the proximate composition (e.g., moisture content, fixed carbon, ash content, total volatile solids, etc.), lignocellulosic composition (e.g., hemicellulose, cellulose, and lignin), biochemical composition as well as other physical and chemical properties of waste materials such as molecular weight (MW), functional group, pH, acidity, calorific value (or heating value), glucans and carbohydrate polymer

[10] Ultimate or elemental analysis constitutes the potency of fuel in the waste materials or biomass.

content (polysaccharide), phenolic polymer content (e.g., lignin), protein content, lipids (or fats and oils), etc., play significant roles in the yields and quality of products formed after thermochemical conversion.

Cellulose[11] is a homopolysaccharide made up of over 3000 β-glycosidic bonds like β-D-glucopyranose[12] units (Haldar and Purkait 2020). Cellulose serves as the source of sugars in waste materials during thermochemical conversion. High cellulose content in waste materials gives rise to high-quality bioproducts with high yields and high technological efficiency during thermochemical conversion processes like pyrolysis, gasification, combustion, etc. (see Chap. 3 for details). Cellulose in waste materials can result in about 11–48% levoglucosan during thermochemical conversion technologies depending on the lignin content (Padmanathan et al. 2024). However, cross-linked cellulose in lignin carbohydrate complex (LCC) may show a unique chemical characteristic compared to pure cellulose because of the formation of C6 position in the chemical structure that can hinder levoglucosan formation (Patwardhan et al. 2009).

Hemicellulose is a polysaccharide in lignocellulosic materials which can easily be depolymerized[13] compared to cellulose thus increasing the sugar content of the waste material during thermochemical conversion. High hemicellulose waste materials usually go through thermal decomposition at comparatively lower temperatures than those with high cellulose content, thus improving products' yields and quality.

Lignin is a complex polymer that makes the cell walls of plants rigid by giving it structural support. Lignin is the most abundant source of aromatic biomolecules and is considered more thermally stable. Nonetheless, waste materials with high quantities of lignin may hinder efficient thermochemical conversion by reducing products' yield due to the rigid nature of lignin making it less susceptible to thermal decomposition into biofuels. Hence, high lignin waste materials are the best source of feedstock for biochar production. Nonetheless, lignin is also found to hinder the activities of catalysts by causing them to be deactivated during thermochemical conversion (Padmanathan et al. 2024). Thus, waste materials and biomass with lower lignin content are favorable for efficient thermochemical conversion into fuels and chemicals. However, because the chemical valency[14] of the carbon atoms in lignin molecules is lower than that in cellulose and hemicellulose, lignin is considered to have higher energy content than cellulose and hemicellulose. The presence of lignin in biomass gives rise to compounds like phenols, furans, and levoglucosan during thermal treatment (Padmanathan et al. 2024).

The chemical composition of feedstocks establishes the extent to which the thermochemical conversion process proceeds with a high rate of conversion and efficiency. The reactivity of other chemical reactants with the chemical components

---

[11] Cellulose is the primary component of plant cell walls comprising of complex carbohydrates. It is a significant source of carbon.

[12] Glucopyranose is a building block or structural component of plant cell wall.

[13] Depolymerization is the process of decomposing polymeric chains to release monomers.

[14] Valency is a property of a given element which determines the number of chemical bonds other atoms will form with an atom of the given element.

of different feedstocks, especially waste materials which comprise a wide range of materials may change due to the varying complex compounds present. For instance, due to the presence of aldehydes or ketones, the reactivity of polysaccharides such as carbohydrates with oxygen is high (oxidation reaction) during a thermochemical conversion process may vary compared to the reaction with protein components in the waste material.

During thermochemical conversion processes, heat is mostly added to the reaction to generate oil, gas, and char. Thus, the reaction of heat with organic acidic components in the waste materials results in the decomposition of these organic acids into simpler compounds such as carbon dioxide, water, and other volatile compounds. Hence, depending on the targeted bioproducts during thermochemical conversion, the acidic organic compounds in the feedstock play a very important role in the yield and quality of specific products generated. Volatile matter content of waste materials may affect the biofuel reactivity which also impacts the yield of fuel. Waste materials with volatile matter of more than 40% are best for thermochemical conversion into fuels and chemicals (CPHEEO 2000).

When waste material contains high protein content, the amino acids (which are the building blocks of proteins) combine with the heat through a series of reactions such as decarboxylation, and the products generated (i.e., amines) depend on the concentration of amino acids. On the other hand, during thermochemical conversion of waste materials with high contents of fatty acids (such as food or animal wastes), they undergo thermal cracking where the triglyceride molecules break down into complex mixtures of bioliquids, biogas, and biochar consisting of hydrocarbons, ketones, aldehydes, paraffins, olefins, etc. (Padmanathan et al. 2024).

The type of functional groups in biomass structure also contributes greatly to the yield of bioproducts during thermochemical conversion. For instance, the presence of hydroxyl (–OH) groups in carbohydrates facilitates the dehydration reactions hence increased the formation of oils in the biomass (Mayes and Broadbelt 2012; Yan et al. 2010; Stankovikj et al. 2016). Nonetheless, higher number of hydroxyl groups in the biomass' carbohydrate structure can also lead to the formation of water during thermochemical conversion process thus decreasing the formation of oil (Parihar et al. 2017).

The moisture content of waste materials affects the overall energy efficiency of the conversion process thus subsequently impacting the energy requirements and combustion efficiency of the biofuel or bioproduct generated. Waste materials with high moisture content above 45% may need additional energy to dry them before the conversion process to achieve high product and technological efficiencies (CPHEEO 2000). Moisture content can affect the yield of products from thermochemical conversion.

Apart from product yield and quality, the knowledge of waste materials' chemistry determines energy efficiency and combustion efficiency. The ultimate or elemental composition such as Carbon (C), oxygen (O), nitrogen (N), hydrogen (H), sulfur (S), etc., also influence the efficiency of thermochemical conversion though they may be found in small quantities in waste materials. Depending on the type of thermochemical conversion technology, the carbon contents in waste materials may affect the yield of their bioproducts in different ways. For instance, high carbon content waste materials (such as biomass with high lignin content) are advantageous in gasification to attain high syngas (or synthesis gas) yield compared to other thermochemical conversion technologies (Yan et al. 2010). Again, high fixed carbon and volatile matter content in waste materials make them highly organic and excellent energy carriers which are excellent properties of raw materials for thermochemical conversion into bioenergy and biofuels. The elemental composition of waste materials also affects the quality of ash and char generated during thermochemical conversion. Again, the composition of syngas[15] hence its usability depends on the elemental composition of waste materials during thermal decomposition. On the other hand, compounds such as carbon, sulfur, and depending metals in feedstock may contribute to emissions of carbon dioxide ($CO_2$), Sulphur dioxide ($SO_x$), acid rain, nitrogen oxides ($NO_x$), ozone, etc., where the impact of these emissions depend highly on the amount of these chemical compounds in the waste material or feedstock.

Minerals and other inorganic compounds in waste materials contribute to their ash content which intends to affect the yield of bioproducts and process efficiency (due to corrosion and fouling in equipment) during thermochemical conversion. High content of ash in waste materials may contribute to corrosion, equipment fouling, and other issues during thermochemical conversion.

To achieve the multiple benefits of agricultural waste (waste management, sustainability, etc.), all the categories of this waste have specific characteristics and components which make them suitable for efficient utilization for thermochemical conversion into biofuels, bioenergy, and biochemicals.

### *1.3.1 Chemistry of Agricultural Wastes and Thermochemical Conversion*

Depending on optimum conditions of thermochemical conversion, dried agricultural waste can generate up to about 80% biofuel yield (Bharathiraja et al. 2018). Generally, based on the type and location of generation or collection, agricultural waste materials are about 75–90 wt% lignocellulosic which consists mainly of about 35–55 wt% cellulose, 15–25 wt% lignin, 25–40 wt% hemicellulose (Ziaei-Rad et al. 2021; Ates et al. 2020). Other compositions include moisture (about 5–10 wt%), ash, hydrocarbons, proteins, some lipids, and other chemical compounds.

Agricultural wastes with high cellulose and hemicellulose contents such as wheat straw and switchgrass were found to release high amount of syngas whilst high lignin agricultural wastes like eastern redcedar produced significantly high methane concentrations upon thermochemical conversion (Pasangulapati et al. 2012). Mixture

[15] Syngas, also called producer gas, town gas, wood gas, raw gas, or synthetic gas is a mixture of gases (mainly 30–0% CO, 25–30% $H_2$, 0–5% $CH_4$, 5–15% $CO_2$ and traces of water vapor and sulfur) produced by thermal decomposition of organic materials at low to high temperatures depending on the technology of waste conversion (Yan et al. 2010).

of agricultural residues as well as rice bran, olive residues, and rice husks have high ash content (18–23%) below the recommended level (<10%) which may pose fouling and corrosion issues with equipment during thermochemical conversion (Zaini et al. 2019).

The carbon content[16] in agricultural wastes is the main source of energy during thermochemical conversion process. Hence, the yield and quality of target product formed depend largely on the carbon content of the raw material or biomass. High carbon contents above 40 wt% in agricultural wastes are found to generate high-quality gaseous bioproducts via thermochemical conversion (Yin 2011). For instance, during gasification of agricultural waste, the high carbon content materials are converted into high-quality syngas (with high hydrocarbon content) at high yields with high heating value. On the other hand, during combustion, for instance, agricultural wastes that have low carbon content like orange peels (about 16–17 wt%) may have negative impact on the combustion efficiency and heat release rate (Yin 2011). The Central Public Health and Environmental Engineering Organization (CPHEEO) concluded that for efficient generation of thermochemical conversion bioproducts, total carbon less than 35% is recommended (CPHEEO 2000). Table 1.1 summarizes the ultimate and elemental composition of agricultural waste materials compared to conventional coal.

Nonetheless, the mix of other elemental components like nitrogen (N), oxygen (O), etc., in waste materials also affects the quality of products, efficiency of thermochemical conversion as well as product yield. For instance, a high carbon-to-nitrogen (C/N) ratio of agricultural wastes is found to cause incomplete combustion resulting in the evolution of particulate matter into the environment, as well as generating poor quality char and ash (Dhakal et al. 2022).

Agricultural wastes such as forest residue, pine chips, corn straw, oak wood, etc., have high heat content which facilitates the extraction of energy from them via thermochemical conversion.

However, because different agricultural wastes have different chemical and physical characteristics, the yield and quality of bioproducts generated after thermochemical conversion may also vary.

Agricultural wastes such as groundnut shells and corn cobs were examined for variations in their moisture content impact on heating value after thermochemical conversion. For instance, high heating values of about 16–21 MJ/kg and 39–43 MJ/kg were achieved with moisture contents of about 8.9–9.3% and 7.5–8.3%, respectively (Burubai and Okpala 2017). However, the increased moisture content (10–24%) of these agricultural waste materials drastically reduced their heating value by about 46–60% (Burubai and Okpala 2017). Nonetheless, some types of agricultural wastes such as palm kernel shells and coconut shells were found to possess the highest heating values at all the tested elevated moisture contents thus these wastes can be cost-effective sources of feedstock for thermochemical conversion into fuels and chemicals. Agricultural wastes like wheat straw, oat straw, and rice straw also have

[16] The carbon content of agricultural wastes varies depending on the type of waste. However, all agricultural waste materials have carbon content ranging from 30 to 45 wt% (Yin 2011).

**Table 1.1** Proximate and ultimate analysis of agricultural waste materials (Yin 2011; Zaini et al. 2019)

| Agricultural waste | Elemental/Ultimate analysis (wt%) | | | | | Proximate analysis (wt%) | | | | High Heating value (MJ/kg) |
|---|---|---|---|---|---|---|---|---|---|---|
| | C | H | N | S | O | Volatiles | Fixed carbon | Ash | Moisture | |
| Conventional[17] coal | 45–60 | 4–6 | 0.4–2 | 0.3–1.5 | 20–25 | 15–30 | 35–55 | 5–45 | 6–32 | 23–27 |
| Rice husks[18] | 39–50 | 4–6 | 0.7–2 | 0–0.5 | 35–53.7 | 61–74 | 11–16 | 15–23.2 | 10 | 13–15 |
| Rice Straw[19] | 34–45 | 4–7 | 0.2–0.8 | 0.1–0.2 | 32–42 | 71–93 | 14 | 8–23 | 4–6 | 14–17 |
| Rice bran | 38.7 | 5.4 | 0.21 | – | 0.67 | 64.3 | 9.2 | 18.8 | 7.7 | 13.4 |
| Corn Stover[20] | 35–46 | 5–7 | 0.3–0.8 | 0.1–0.3 | 43–46 | 4–7 | 5–8 | 4–7 | 5–8 | 16–17 |
| Corn cobs[21] | 40–53 | 5–6 | 0.05–0.06 | 0.01–0.03 | 45–46 | 61–71 | 17–27 | 3–4 | 3–4 | 18–19 |
| Corn straw[22] | 44– | 5.9 | 0.6 | 0.07 | 40– | 73.2 | 19.2 | 5–6 | 15–20 | 17–18 |
| Hazelnut shell | 54.8 | 6.7 | 1.0 | 0.1 | 37.4 | 72 | 21 | 7 | 7–10 | 18.2 |
| Sugarcane bagasse | 40–46 | 5.9 | 0.15 | – | 45.2 | 79–84 | – | 3–5 | 16.1 | 18–19 |
| Olive residue | 48.3 | 5 | 1.2 | 0 | 45.5 | 68.6 | 10.4 | 21 | 7–8 | 15.9 |
| Spent coffee grounds[23] | 45–50 | 6–7 | <2 | <0.4 | 40.9 | 97–99 | – | <1 | 10–15 | 18–25 |

(continued)

[17] Johar et al. (2023)

[18] Biswas et al. (2017), Kim and Lim (2022), and Yin (2011)

[19] Kumar et al. (2018)

[20] Kumar et al. (2018)

[21] Yin (2011)

[22] Yin (2011)

[23] Malagón-Romero et al. (2023) and Czekała et al. (2023)

**Table 1.1** (continued)

| Agricultural waste | Elemental/Ultimate analysis (wt%) | | | | | Proximate analysis (wt%) | | | | High Heating value (MJ/kg) |
|---|---|---|---|---|---|---|---|---|---|---|
| | C | H | N | S | O | Volatiles | Fixed carbon | Ash | Moisture | |
| Coffee husks[24] | 47– | 6.4 | – | 43.7 | – | 78.5 | 19.1 | 2.4 | – | 19.8 |
| Sunflower seed shells | 47.8 | 6.1 | 1.2 | 0.3 | 44.6 | 83.7 | 13.6 | 2.7 | 6–7 | 17.7 |
| Peanut shells | 45–50 | 6.1 | 2.1 | – | 44.4 | 84.9 | 13.4 | 1.7 | 8–45 | 18.6 |
| Wheat straw[25] | 40–47 | 5–6.5 | 0.4–0.6 | 0.1–0.3 | 34–52 | 74–93 | 17.3 | 7–13 | 4–9 | 17–19 |
| Oat straw | 48 | 6 | 0.5 | 0.08 | 44.6 | 74.5 | 19.5 | 5.9 | 5.4 | – |
| Barley straw[26] | 49 | 6.2 | 0.7 | 0.13 | 44.6 | 76.2 | 19.8 | 5–10 | – | 16.4 |
| Oil palm empty fruit bunches | 45–49 | 5–8 | 0–1.2 | 0.6–1.1 | 40–48 | 71–80 | 8–19 | 3–8 | 67 | 16–19 |
| Orange peels | 16.2 | 17.1 | 0.8 | 0.76 | 60.2 | – | – | – | – | 17–29 |
| Coconut shell | 50 | 5.7 | 43.4 | – | – | 77.2 | 22.1 | 0.7 | 7–12 | 20–35 |
| Palm kernel shell[27] | 88–99 | 1–14 | 0.03–0.14 | 0.01–0.23 | | 57–64 | 20–37 | | 3–12 | 38.3 |
| Bamboo wood | 48.6 | 6.3 | 0.2 | – | 42.8 | 86.8 | 11.2 | 1.9 | 6–15 | 20.6 |
| Cotton stalk | 47 | 4.6 | 1.15 | – | 42.1 | 76.1 | 18.8 | 5.1 | 8–15 | 17.4 |
| Cotton residue | 47 | 5.9 | 1.8 | 0.19 | 38.4 | 72.8 | 20.6 | 6.6 | – | 16.9 |
| Forest residue | 53– | 6.3 | 0.3 | 0.09 | 40 | 79.8 | 20.0 | 0.2 | – | 19.5 |

(continued)

[24] Yin (2011)

[25] Kumar et al. (2018), Yin (2011), and Pasangulapati et al. (2012)

[26] Serrano et al. (2011)

[27] Mogaji (2020)

**Table 1.1** (continued)

| Agricultural waste | Elemental/Ultimate analysis (wt%) | | | | | Proximate analysis (wt%) | | | | High Heating value (MJ/kg) |
|---|---|---|---|---|---|---|---|---|---|---|
| | C | H | N | S | O | Volatiles | Fixed carbon | Ash | Moisture | |
| Agric waste/residue mix[28] | 49–51 | 5–6 | 2.4 | 0.7 | 28.9 | 69 | 18.1 | 12.5 | – | 18.4 |
| Switch grass | 46.7 | 5.7–5.9 | 0.1–0.7 | 0.19–0.3 | 37–43 | 76.7 | 14.3 | 8.9 | <15 | 18.1 |
| Oak wood branches | 48–49 | 6–7 | 2–3 | – | 42–42.3 | 77–82 | 16–19 | 2–4.1 | 8–14 | 19–19.3 |
| Pine chips | 49–50 | 5.7 | 0.5 | 0.08 | 38.07 | 72.4 | 21.7 | 5.9 | – | 19.8 |
| Animal manure[29] | 18–44 | – | 12–22 | – | 34.4 | 75 | 10–15 | 20–50 | | 4.5–20 |
| Spent mushroom substrate[30] | 29–36 | – | 1.69–1.8 | – | – | – | – | 39–49 | 53–72 | 13–19 |
| Eastern redcedar[31] | 51.1 | 5.97 | 0.37 | 0 | 40.95 | – | – | – | 12 | 10.2 |

[28] Yin (2011)

[29] Azevedo and Stout (1974)

[30] Dávila et al. (2020)

[31] Pasangulapati et al. (2012)

very low moisture content (about 4–9 wt%) and thus are excellent feedstock for thermochemical conversion into biofuels and biochemicals. The moisture content of agricultural wastes of less than 30% is considered ideal for specific thermochemical conversion technologies such as gasification (Burubai and Okpala 2017).

### *1.3.2 Effects of MSW Chemistry on Thermochemical Conversion*

Table 1.2 shows the elemental composition of municipal solid waste components.

To appropriately select the right thermochemical conversion method, the physicochemical profile of the feedstock must be known because composition significantly affects the yield and components of products generated. Since MSW is made up of different waste components, it is important to know the chemical component distribution and how they are altered during thermochemical conversion.

Thermochemical conversion processes like combustion, pyrolysis, and gasification can cause chemical alterations to the components of MSW by the application of heat to convert the organic and inorganic matter into different products like bio-oil, bio-syngas, char, etc., during thermochemical treatment of MSW, the complex organic compounds are decomposed into simpler ones (e.g., gases, volatiles, solid residues, and liquids) resulting in the reduction of the bulk or volume of waste treated (Nardella et al. 2022). MSW contains metals, glass, and other materials which are recalcitrant to heat and will not be volatilized at elevated temperatures thus the variation in the properties of the solid residues (biochar, ash, etc.) generated. During thermochemical conversion at high temperatures, heavy metals and organic pollutants are degraded, hence obtaining eco-friendly products with little emissions (Yang et al. 2024).

MSW mix contains very high volatile matter (about 50–55%) which is higher than agricultural wastes and conventional coal, thus, MSW can ignite and gasify faster at elevated temperatures (about 15–17 wt% moisture) (Liu et al. 2020). The sulfur content in MSW ranges from about 0.03–0.96 wt% with the main source being hair, rubber, fiber, etc. (Liu et al. 2020). The organic sulfur and chlorine (from food waste, plastics, etc.) in MSW are mostly released at temperatures below 500 °C where they form $SO_2$ and hydrochloric acid (HCl) in the atmosphere (Ma et al. 2022). At elevated temperatures (>800 °C), inorganic chlorine in food waste and biomass could react with organic functional groups or $SiO_2$ and vaporize partially to form HCl (Frandsen et al. 2007).

Plastics as part of MSW mix act as a hydrogen donor to achieve hydrodeoxygenation of products hence obtaining high synergistic gains from other components in MSW mix (Wang et al. 2024). The synergistic effect of pyrolyzing plastics together with other municipal solid waste is normally controlled by the interactions of free

**Table 1.2** Proximate and ultimate analysis of MSW composition (Ibrahim et al. 2020)

| Municipal Solid Waste[32] | Elemental/Ultimate analysis (wt%) | | | | | Proximate analysis (wt%) | | | | High heating value (MJ/kg) |
|---|---|---|---|---|---|---|---|---|---|---|
| | C | H | N | S | O | Volatile matter | Fixed carbon | Ash | Moisture | |
| Paper | 43.5 | 6 | 0.3 | 0.2 | 44 | 12–14 | 16–17 | – | – | 10–27 |
| Plastics | 60–69 | 7.2 | – | – | 22.8 | 90 | – | 0.01 | – | 19–47 |
| Textiles | 55 | 6.6 | 4.6 | 0.2 | 31.2 | – | – | – | – | 15–29 |
| Glass | 0.5 | 0.1 | 0.1 | – | 0.4 | – | – | – | – | 18.6 |
| Metal | 4.5 | 0.6 | – | – | 4.3 | – | – | – | – | – |
| MSW mix | 20–53.8 | 5–7 | 0.39–1.7 | 0.2–0.9 | 32.9–48 | 55.1 | 8–98 | 7–20 | 16.9 | 9–20.93 |

[32] Ibrahim et al. (2020)

radicals obtained from the other MSW components, and the efficiency of this interaction is based on contact time of the volatile components (Yang et al. 2024; Wang et al. 2024).

## 1.4 Global Waste Generation and Consumption Statistics

### *1.4.1 Agricultural Waste Generation and Consumption Profile*

The generation of some types of waste like agricultural wastes is unavoidable due to human activities, and its generation rate is projected to increase with the increase in population growth and living standards. Due to population growth, the world's growth in agricultural crop production in 2021 was very high at about 8 billion tons, and this is estimated to increase by over 50% by the year 2050 (Otsuka and Fan 2021). As a result, there has been a rapid rise in agricultural waste generation with an annual growth rate of about 5–10% leading to environmental pollution (Debnath et al. 2021). About 140 billion metric tons of agricultural waste materials are generated globally every year which mainly includes rice straw, bagasse, wheat straw, and corn straw (Otsuka and Fan 2021; Riseh et al. 2024).

Rice straw is one of the most generated agricultural wastes in the world with over 730 million tons/year generation rate (Sarkar et al. 2012). In 2016, the world's crop residues[33] production was estimated at about 2802 Mt/y with only about 2–10% utilized for biofuels and bioenergy (Kumar et al. 2018). For a total crop residue of about 9667 Mt/y including legumes and other food crops, about $11.3 \times 10^{15}$ kcal (amounting to about 7.5 billion barrels, bbl) of biodiesel can be produced from them (Lal 2005). Annually, rice straw as the most abundant agricultural waste or crop residue in the world is generated at a rate of about 731 Mt (Sarkar et al. 2012). The United States Department of Energy estimated the total crop residue generation to be about 180 million dry tons by the year 2030 (US DOE 2011). Figure 1.2 shows the global generation of common agricultural crop residues.

### *1.4.2 Municipal Solid and Liquid Waste: Global Statistics*

Due to rapid population growth, the generation of municipal liquid waste such as sewage sludge is concerning, and it is becoming a limitation to sustainable development if not managed properly. In China alone, for example, about 76.763 million tons

[33] This amount is estimated for only cereal crops. For legumes and other food crops, the total crop residue generation in the world is about 9667 Mt/y (Lal 2005).

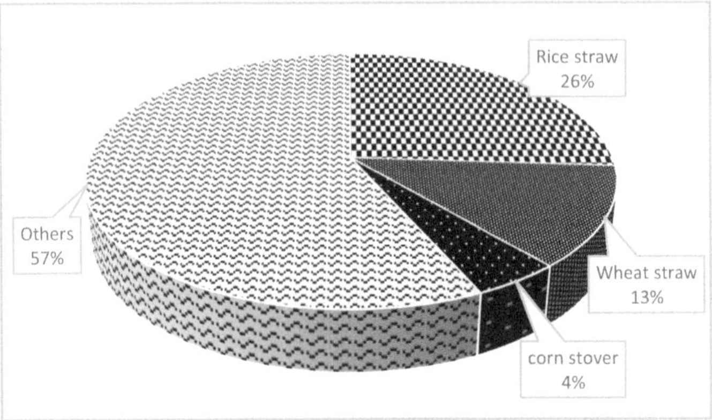


**Fig. 1.2** Global generation of major components of agricultural wastes (US DOE 2011)

of sludge was generated in 2021 with about 1.21% average annual increase (Liew et al. 2022).

Globally, MSW generation is estimated at about 2 billion tons annually (Zhao and Li 2023) with a projection of about 3.5 billion annually by the year 2050 (Munawar et al. 2023; Kaza et al. 2018; Zhao and Li 2023). About 0.7 billion tons of the MSW generated annually is not collected and recycled by municipal authorities (Zhao and Li 2023). For instance, in 2018, only about 32% of the total generated MSW was recycled and treated as compost and fertilizers (EPA 2018). For the total recycled MSW, less than 42% was usually utilized for bioenergy and biofuels (EPA 2018). In 2021, thermochemical conversion technologies such as combustion, were used to reduce about 28 million tons of MSW by generating about 13.6 billion kWh of electricity (US EIA 2022; OECD 2021). According to the World Health Organization (WHO), about 97% of cities in rural and urban areas dispose of MSW in landfills without proper waste management (Kaza et al. 2018). During the natural decomposition of MSW in landfills and incineration sites, $CO_2$ and $CH_4$ are mostly emitted which contributes about 70% of greenhouse gas emissions (Kaza et al. 2018). Figures 1.3a and 1.3b show the share of MSW compositions in the utilization of energy via thermochemical conversion technologies and their GHG benefits to the environment.

Annually, over 50% of the total generated MWS was disposed of into over 1278 landfill sites which pose global economic and environmental burdens (OECD 2021). In 2018, about 2.5 million tons of paper, plastic plates, and cups, and about 95% of food waste from municipalities were disposed of without efficient utilization (OECD 2021). In 2018, about 9.2 million tons of rubber, leather, and textiles were generated where only about 1.7 million tons which were rubber from tires were recycled (EPA 2018). About 27% of the total rubber, leather, and textiles generated annually is thermally decomposed via combustion or other thermochemical conversion technologies whilst over 50% are disposed of in landfills (OECD 2021). Food waste and yard trimmings have been effectively managed by few municipalities for organic fertilizers

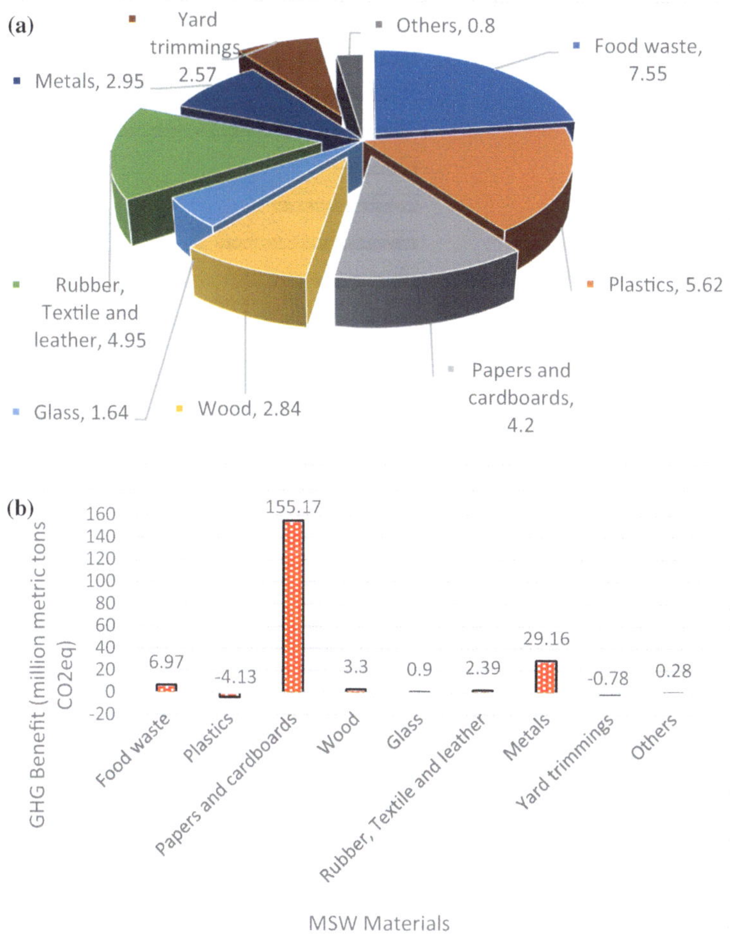


**Fig. 1.3** **a** Amount of MSW used as feedstock for bioenergy recovery via combustion (OECD 2021). **b** Greenhouse gas benefits of MSW recycling, composting, and combustion for bioenergy (OECD 2021)

instead of usual uncontrolled landfilling and incineration. In 2018, about 58–60% of generated food waste and yard trimmings were used as compost (EPA 2018). Annually, MSW generation per capita increases by about 9% per person per day (EPA 2018). Due to technological advancement, the generation of paper and paperboard waste has consistently declined by about 18–23% since the year 2000 (EPA 2018). Figure 1.4 shows the generation and utilization (recycling, reuse, composting, and combustion) of MSW in 2018 (EPA 2018).

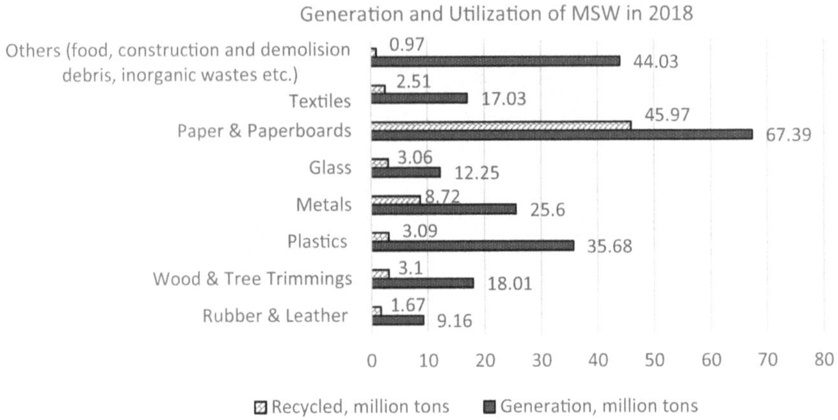


**Fig. 1.4** Composition of municipal solid waste generated and recycled in 2018 (EPA 2018)

## 1.5 Conclusion

Due to global population growth, the generation of waste has increased over the years. Every year about 80–90% of the total global waste generated is disposed into the environment with proper management thus posing environmental concerns. Nonetheless, the waste generated has an extensive diversity of carbonaceous waste materials such as municipal solid wastes (MSW), industrial hazardous and non-hazardous wastes, and agricultural residues (e.g., lignocellulosic biomass) which are excellent sources of raw materials for cost-efficient production of biochemicals, biofuel, and bioenergy through thermochemical conversion technologies. The types of biomass wastes and other potential waste materials are characterized predominantly by their physicochemical properties which eventually determine the quality, calorific value, and quantity of bioproducts generated via thermochemical conversion. The chemistry of waste components has synergistic effect on the products during thermochemical conversion processes. Thus, the conversion of waste materials would be efficient when the suitable type of waste is converted via the appropriate thermochemical conversion technology. The major potential carbonaceous feedstocks/waste that are outstanding for thermochemical conversion into bioproducts were discussed in this chapter.

## References

Ates B, Koytepe S, Ulu A, Gurses C, Thakur VK (2020) Chemistry, structures, and advanced applications of nanocomposites from biorenewable resources. Chem Rev 120:9304–9362

Azevedo J, Stout PR (1974) Farm animal manures: an overview of their role in the agricultural environment. Manual 44. Division of Agricultural Sciences, University of California. Berkeley, CA

Beder S (1996) The nature of sustainable development, 2nd edn. Scribe Publications, Newham

Bharathiraja B, Sudharsana T, Jayamuthunagai J, Ramanujam PK (2018) Biogas production—a review on composition, fuel properties, feed stock and principles of anaerobic digestion. Renew Sustain Energy Rev 90:570–582

Biswas B, Pandey N, Bisht Y, Singh R, Kumar J, Bhaskar T (2017) Pyrolysis of agricultural biomass residues: comparative study of corn cob, wheat straw, rice straw and rice husk. Bioresour Technol 237:57–63

Bodzay B, Banhegyi G (2017) Polymer waste: controlled breakdown or recycling? Int J Design Sci Technol 22(2):109–138

Burubai W, Okpala AN (2017) Effect of moisture on the thermal capacity of some agricultural wastes. Int J Eng Res Dev 13(1):32–36

Central Public Health and Environmental Engineering Organization (CPHEEO) (2000). Energy recovery from municipal solid wastes, manual on solid waste management. https://19january2021snapshot.epa.gov/smm/energy-recovery-combustion-municipal-solid-waste-msw_.html#:~:text=Energy%20recovery%20from%20waste%20is%20the%20conversion%20of,This%20process%20is%20often%20called%20waste%20to%20energy. Accessed 19 Dec 2024

Czekała W, Łukomska A, Pulka J, Bojarski W, Pochwatka P, Kowalczyk-Juśko A, Oniszczuk A, Dach J (2023) Waste-to-energy: biogas potential of waste from coffee production and consumption. Energy 276:127604

Dávila LRG, Walter MA, Zambrano CJF, Héctor SM, Méndez JJA (2020) Evaluation of nutritional values of wild mushrooms and spent substrate of Lentinus crinitus (L.) Fr. Heliyon 6(3):e03502

Debnath B, Haldar D, Purkait MK (2021) A critical review on the techniques used for the synthesis and applications of crystalline cellulose derived from agricultural wastes and forest residues. Carbohydr Polym 273:118537

DEFRA (Department for Environment, Food, and Rural Affair) (2012) UK government department for policy and regulation on the environment, food and rural affairs. http://www.defra.gov.uk/. Accessed 2 Mar 2024

Department of Energy and Environmental Protection (DEEP) (2016). Construction and demolition waste characterization and market analysis. https://portal.ct.gov/deep. Accessed 14 Mar 2024.

Dhakal S, Minx JC, Toth FL, Abdel-Aziz A, Figueroa Meza MJ, Hubacek K, Jonckheere IGC, YongGun K, Nemet GF, Pachauri S, Tan XC, Wiedmann T (2022) Emissions trends and drivers. In: Shukla PR, Skea J, Slade R, Al Khourdajie A, van Diemen R, McCollum D, Pathak M, Some S, Vyas P, Fradera R, Belkacemi M, Hasija A, Lisboa G, Luz S, Malley J (eds.) IPCC, 2022: climate change 2022—mitigation of climate change. Contribution of working group III to the sixth assessment report of the intergovernmental panel on climate change, Cambridge University Press, Cambridge

Environmental Protection Agency (EPA) (2018) National overview: facts and figures on materials, wastes and recycling. https://www.epa.gov/facts-and-figures-about-materials-waste-and-recycling. Accessed 5 Jul 2022.

Frandsen F J, van Lith SC, Korbee R, Yrjas P, Backman R, Obernberger I, Brunner T, Jöller M (2007) Quantification of the release of inorganic elements from biofuels. Fuel Process Technol 88 (11–12):1118–1128.

Haldar D, Purkait MK (2020) Micro and nanocrystalline cellulose derivatives of lignocellulosic biomass: a review on synthesis, applications and advancements. Carbohydr Polym 250:116937

Hale RC, La Guardia MJ, Harvey E, Chen D, Mainor TM, Luellen DR, Hundal LS (2012) Polybrominated diphenyl ethers in U.S. sewage sludges and biosolids: temporal and geographical trends and uptake by corn following land application. Environ Sci Technol 46(4):2055–2063

Hernández LL, Zeeman G, Temmink H, Buisman C (2007) Characterization and biological treatment of greywater. Water Sci Technol 56(5):193–200

Ibrahim N, Abo-el-Ata G, El-Hattab M (2020) Status, problems and challenges for municipal solid waste management in Assiut Governate. J Environ Stud Res 10(2):362–384

International Energy Agency (IEA). (2021). Transport biofuels. https://www.iea.org/reports/renewables-2020/transport-biofuels. Accessed 25 Jan 2023.

Johar P, Rylott EL, Robert McElroy C, Matharu AS, Clark JH (2023) Biologically bound nickel accelerated de-polymerization of polyethylene to high value hydrocarbons and hydrogen. RSC Sustainability 1(1):117–127

Kariim I, Swai H, Kivevele T (2022) Recent advances in thermochemical conversion of biomass into drop-in fuel: a review. Sci Afr 17:e01352

Kaza S, Yao LC, Bhada TP, Van WF, et al (2018) What a waste 2.0: a global snapshot of solid waste management to 2050. Urban Development Series. Washington, D.C.: World Bank Group. https://documents.worldbank.org/en/publication/documents-reports/documentdetail/697271544470229584/what-a-waste-2-0-a-global-snapshot-of-solid-waste-management-to-2050. Accessed 11 Apr 2023

Kim S, Lim Y-I (2022) Economic evaluation and analysis of electricity and nano-porous silica productions from rice husk. Comput Aided Chem Eng 49:97–102

Kumar S, Paritosh K, Pareek N, Chawade A, Vivekanand V (2018) De-construction of major Indian cereal crop residues through chemical pretreatment for improved biogas production: an overview. Renew Sustain Energy Rev 90:160–170

Lal R (2005) World crop residues production and implications of its use as a biofuel. Environ Int 31(4):575–584

Lee KT, Ofori-Boateng C (2013) Sustainability of biofuel production from oil palm biomass. Springer Science+Business Media, Singapore, p 323

Letshwenyo MW, Kgetseyamore D (2020) Generation and composition of municipal solid waste: case study, extension 7, Palapye, Botswana. SN Appl Sci 2:1665

Li P, Zhao H, Cheng C, Hou T, Shen D, Jiao Y (2024) A review on anaerobic co-digestion of sewage sludge with other organic wastes for methane production: mechanism, process, improvement and industrial application. Biomass Bioenerg 185:107241

Liew CS, Yunus NM, Chidi BS, Lam MK, Goh PS, Mohamad M, Sin JC, Lam SM, Lim JW, Lam SS (2022) A review on recent disposal of hazardous sewage sludge via anaerobic digestion and novel composting. J Hazard 423:126995

Liu H, Wang Y, Zhao S, Hu H, Cao C, Li A, Yu Y, Yao H (2020) Review on the current status of the co-combustion technology of Organic Solid Waste (OSW) and coal in China. Energy Fuel 34:15448–15487

Lowell Center for Sustainable Production (LCSP) (1998) Sustainable production: a working definition. Informal Meeting of the Committee Members.

Ma W, Liu X, Ma C, Gu T, Chen G (2022) BASIC: a comprehensive model for SOx formation mechanism and optimization in municipal solid waste (MSW) combustion. ACS Omega 7:3860–3871

Malagón-Romero D, Torres-Velasquez AC, Tinoco-Navarro LK, Arrubla-Vélez JP (2023) Pyrolysis of Colombian spent coffee grounds (SCGs), characterization of bio-oil, and study of its antioxidant properties. Int J Sustain Energy 42(1):811–829

Mayes HB, Broadbelt LJ (2012) Unraveling the reactions that unravel cellulose. J Phys Chem A 116:7098–7106

Mogaji TS (2020) Thermal degradation conditions effects on selected biomass wastes and characterization of their produced biochar. J Energy Res Reviews 4(3):46–59

Morel A, Diener S (2006) Greywater management in low and middle-income countries. Swiss Federal Institute of Aquatic Science and Technology (Eawag), Dubendorf

Munawar MA, Li P, Ma Q, Haque MA, Chen W-T (2023) Thermochemical conversions of municipal solid waste into fuels and chemicals. In: Li Y, Chang C (eds) Advances in bioenergy, vol 8. Elsevier, pp 239–305

Nardella F, Bellavia S, Mattonai M, Ribechini E (2022) Co-pyrolysis of biomass and plastic: synergistic effects and estimation of elemental composition of pyrolysis oil by analytical pyrolysis-gas chromatography/mass spectrometry. Bioresour Technol 354:127170

Ofori-Boateng C (2022) Algal biorefining: resource expenditure and exergo-environmental sustainability. Nova Science Publishers, New York, p 296

Organization for Economic Co-operation and Development (OECD) (2021). Municipal waste, generation, and treatment. https://data.oecd.org/waste/municipal-waste.htm. Accessed 11 Oct 2023

Organization for Economic Co-operation and Development-Food and Agriculture Organization (OECD-FAO) (2015) Agricultural Outlook 2015. https://www.oecd-ilibrary.org/agriculture-and-food/oecd-fao-agricultural-outlook-2015/per-capita-meat-consumed-in-the-world_agr_outlook-2015-graph94-en. Accessed 10 Jun 2023

Otsuka K, Fan S (2021) Agricultural development: new perspectives in a changing world. International Food Policy Research Institute, Washington, DC

Padmanathan AMD, Beck S, Ansari KB, Mushrif SH (2024) Impact of lignin–carbohydrate complex (LCC) linkages on cellulose pyrolysis chemistry. Electronic supplementary information (ESI) available: First principles computational methods, images of thin-films preparation used in experiments, comparison of activation barriers in LCC conformers, images of TS configurations, first principles barriers calculated using DFT and atomic coordinates of reaction intermediates. Energy Adv 3(2):515–528

Parihar A, Sripada P, Bambery K, Garnier G, Bhattacharya S (2017) Investigation of functional group changes in biomass during slow pyrolysis using synchrotron based infra-red microspectroscopy and thermogravimetry-infra-red spectroscopy. J Anal Appl Pyrolysis 127:394–401

Pasangulapati V, Ramachandriya KD, Kumar A, Wilkins MR, Jones CL, Huhnke RL (2012) Effects of cellulose, hemicellulose and lignin on thermochemical conversion characteristics of the selected biomass. Bioresour Technol 114:663–669

Patwardhan PR, Satrio JA, Brown RC, Shanks BH (2009) Product distribution from fast pyrolysis of glucose-based carbohydrates. J Anal Appl Pyrolysis 86:323–330

Paulo PL, Azevedo C, Begosso L, Galbiati AF, Boncz MA (2013) Natural systems treating greywater and blackwater on-site: Integrating treatment, reuse and landscaping. Ecol Eng 50:95–100

Riseh RS, Vazvani MG, Hassanisaadi M, Thakur VK (2024) Agricultural wastes: a practical and potential source for the isolation and preparation of cellulose and application in agriculture and different industries. Ind Crops Prod 208:117904

Sahay A (2023) Industrial waste: Management, causes, effects. https://www.embibe.com/exams/industrial-waste/. Accessed 18 Aug 2023

Sarkar N, Ghosh SK, Bannerjee S, Aikat K (2012) Bioethanol production from agricultural wastes: an overview. Renew Energy 37(1):19–27

Serrano C, Monedero E, Lapuerta M, Portero H (2011) Effect of moisture content, particle size and pine addition on quality parameters of barley straw pellets. Fuel Process Technol 92(3):699–706

Smil V (1999) Crop residues: agriculture's largest harvest—Crop residues incorporate more than half of the world's agricultural phytomass. J BioSci 49(4):299–308

Srivastava N, Singh R, Singh P, Ahmad I, Singh RP, Rai AK, Asiri M, Gupta VK (2023) Recent advances on lignocellulosic bioresources and their valorization in biofuels production: challenges and viability assessment. Environ Technol Innov 29:103037

Stankovikj F, McDonald AG, Helms GL, Garcia-Perez M (2016) Quantification of bio-oil functional groups and evidences of the presence of pyrolytic humins. Energy Fuel 30(8):6505–6524

U.S. Department of Energy (2011) U.S. billion-ton update: biomass supply for a bioenergy and bioproducts industry. In: Perlack RD and Stokes BJ (Leads), ORNL/TM-2011/224. Oak Ridge National Laboratory, Oak Ridge, TN, p 227. https://www1.eere.energy.gov/bioenergy/pdfs/btu_crop_residues.pdf. Accessed 10 April 2024

United States Energy Information Administration (EIA) (2022) Waste-to-energy (Municipal solid waste). https://www.eia.gov/energyexplained/biomass/waste-to-energy.php#:~:text=Waste%2Dto%2Denergy%20plants%20make%20steam%20and%20electricity&text=In%202022%2C%2063%20U.S.%20power,combustible%20MSW%20for%20electricity%20generation. Accessed 16 Nov 2023

Wang R, Diao R, Wang C, Zhu X (2024) Insights into co-pyrolytic behavior, emission, kinetics, synergy of bio-oil distillation residue and waxy product via TG-FTIR-MS. Fuel 357(Part B):129853

Wei W, Cai Z, Fu J, Xie G-J, Li A, Zhou X, Ni B-J, Wang D, Wang Q (2018) Zero valent iron enhances methane production from primary sludge in anaerobic digestion. Chem Eng J 351:1159–1165

Yan F, Luo SY, Hu ZQ, Xiao B, Cheng G (2010) Hydrogen-rich gas production by steam gasification of char from biomass fast pyrolysis in a fixed-bed reactor: influence of temperature and steam on hydrogen yield and syngas composition. Bioresour Technol 101(14):5633–5637

Yang G, Peng P, Guo H, Song H, Li Z (2024) The catalytic pyrolysis of waste polyolefins by zeolite-based catalysts: a critical review on the structure-acidity synergies of catalysts. Polym Degrad Stab 222:110712

Yin C-Y (2011) Prediction of higher heating values of biomass from proximate and ultimate analyses. Fuel 90(3):1128–1132

Zaini IN, López CG, Pretz T, Yang W, Jönsson PG (2019) Characterization of pyrolysis products of high-ash excavated-waste and its char gasification reactivity and kinetics under a steam atmosphere. Waste Manage 97:149–163

Zhao Y, Li H (2023) Understanding municipal solid waste production and diversion factors utilizing deep-learning methods. Util Policy 83:101612

Ziaei-Rad Z, Fooladi J, Pazouki M, Gummadi SN (2021) Lignocellulosic biomass pre-treatment using low-cost ionic liquid for bioethanol production: an economically viable method for wheat straw fractionation. Biomass Bioenergy 151:106140

# Chapter 2
# Global Market Profile of Bioproducts from Thermochemical Conversion Technologies

**Abstract** Bioenergy and biofuels form about 70% of today's global renewable energy, and about 10% of global primary energy supply. Thermochemical conversion technologies such as pyrolysis, hydrothermal liquefaction, gasification, etc., have realized immense breakthroughs in research and development where all kinds of biofuels and bioenergy can efficiently be produced from waste materials such as construction and demolition waste, tires, plastic waste, and lignocellulosic biomass at high yields. Though thermochemical conversion technologies are still under continuous improvement, there are many commercial production plants operating in most parts of the world at high efficiencies. The growth margin of biofuels, bioenergy, and biochemicals worldwide is increasing as their market soar day by day. Despite the market expansion, advancement, and improvement in biofuels, there are still obstacles impacting the overall production, consumption, and market opportunities in the world hence fighting off probable investors and consumers. This chapter assesses the world's production, consumption, market prospects, potential financiers, topographical distribution, and growth forecast of biofuels, bioenergy, and biochemicals generated from waste via thermochemical conversion processes.

**Keywords** Thermochemical conversion · Global biofuel market · Biofuel growth forecast · Waste

## 2.1 Introduction

Fossil fuel is presently the most produced and consumed source of fuel in the world due to its low cost compared to biofuels. As a result of the rapid rise in global population growth of about 0.9–1.1% annually, the demand for energy is exponentially escalating with a projected cumulative growth of about 40–45% from now to 2040 (Doshi et al. 2016). For instance, in 2022, the world's population stood at about 8 billion with energy demand of about 167,000–168,000 Terawatt-hour (TWh) (i.e., about 601–605 exajoules [EJ]) which is estimated to increase by about 4.5% by the year 2050 (Statista 2022a). This increased demand in fossil-based fuels has led to



C. Ofori-Boateng, *Sustainability of Thermochemical Waste Conversion Technologies*,
https://doi.org/10.1007/978-3-031-64342-2_2

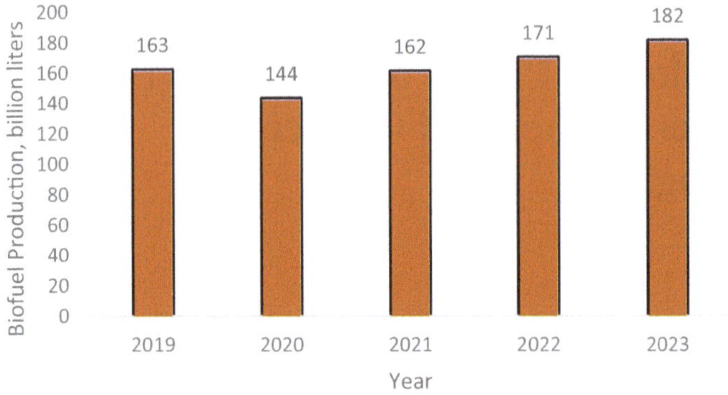


**Fig. 2.1** Annual global biofuel production trend from 2019 to 2023 (IEA 2020)

the contribution of about 70–75% of global greenhouse gas (GHG) emissions which are inimical to living organisms on earth. It is predicted that by 2050, the global rise in $CO_2$ emissions will reach about 43 billion metric tons, from about 37 billion metric tons recorded in 2018 (Statista 2022b). More than 110 countries including the European Union, United Kingdom, South Korea, China, etc., have laid down policies toward reaching net zero carbon target by 2050 where crucial legislation is being implemented to achieve these goals (Chen and Onal 2016).

In view of the effects of fossil fuel use as a menace to the environment as well as its non-renewable nature making it exhaustible in the near future, there has been an insatiable need for sustainable alternative green fuels from waste materials. Though biofuel production rate has been slow over the years, according to the International Energy Agency, the global biofuel production increased from about 163 billion liters to about 183 billion liters in 2023 (IEA 2020) as shown in Fig. 2.1.[1] With the world's aim to have net zero $CO_2$ emissions by 2050, the current annual biofuel production rate of about 7–10 billion liters which corresponds to $CO_2$ emissions reduction of about 10–12 gigaton (Gt) (IEA 2020; Awogbemi and Kallon 2022) is making the vision possible (see Fig. 2.1). Nevertheless, the difficulties in biofuels technological improvement as well as high capital investment and food/fuel discourse can tremendously impact the global production and consumption rate of bioenergy and biofuels negatively.

To offset the economic burdens and facilitate the expansion of the biofuel and bioenergy market worldwide, the provision and implementation of government policies and subsidies to help displace fossil fuels in the near future become necessary. Again, the utilization of waste materials into second-generation biofuels via efficient conversion methods such as thermochemical conversion technologies is a huge way of easing biofuel costs thus expanding its global market potentials.

[1] The decline in 2020 is because of the corona virus pandemic which imposed restrictions and brought most of the biofuel production plants to low operation capacities or total halt.

Thermochemical biomass conversion technologies such as pyrolysis, torrefaction, carbonization, gasification, combustion, hydrothermal liquefaction, etc., can be applied to transform waste materials into green fuels such as bio-oil, bio-gasoline, bio-syngas, bio-methane, biodiesel, etc., which can be used in automobile engines without modification. These biofuels are biodegradable, ecologically friendly, and sustainable especially when derived from waste materials. The expansion of biofuel production could contribute to the generation of employment, community development, social sustainability, and poverty reduction in most parts of the world (Nakamya 2022).

Though the development of these technologies is ongoing, there are a couple of commercialized biofuel production plants in most parts of the world including the United States, Europe, and Asia (Srivastava et al. 2023). This chapter discusses the global market potentials as well as consumption and production rates of common biofuels and bioenergy generated from waste materials using thermochemical conversion technologies.

## 2.2 Global Outlook of Liquid Biofuels from Waste via Thermochemical Conversion

### *2.2.1 Pyrolysis Oil: Global Production and Consumption Profile*

Pyrolysis oil is a dark greenish or brownish organic liquid product obtained from the condensation of condensable gases released during pyrolysis or thermal decomposition of biomass and other waste materials in the absence of oxygen with or without catalysts. Pyrolysis oil can be used as fuel (for heating, power generation, etc.) or as biochemical or additive in the transportation, thermal plants, and pharmaceutical industries. There are numerous active commercial pyrolysis oil production plants in the world that contribute to the reduction of $CO_2$ emissions in the atmosphere by utilizing waste materials to produce sustainable and clean fuels and energy.

According to BTG Bioliquid company,[2] in 2015, commercial waste (sawmill wood waste) pyrolysis plants in the Netherlands (Empyro Hengelo), Finland (Green Fuel Nordic Oy), and Sweden (Pyrocell) with production capacities of about 20 million liters of pyrolysis oil annually, began operation, being the first commercial scale fast pyrolysis oil production plant in the world (IEA 2021). Onym Group, a 6 million liters/year commercial pyrolysis plant also began operation in 2023 where it processes wood waste into pyrolysis oil via fast pyrolysis in an auger reactor (IEA 2021).

[2] BTG Bioliquids (BTG-BTL) is a fast pyrolysis technology provider which employs the rotating cone pyrolysis reactor at high product efficiency.

Ensyn Corporation is presently the world's leading producer of pyrolysis oil whose licensed facilities are in Renfrew and Port Cartier in Canada with a combined operation capacity of about 56 million liters/year (Emery et al. 2021). Ensyn' fast pyrolysis plant supplies bio-crude oil generated from at least about 65,000[3] dry metric tons of forest residues per year as fuel and biochemicals to mainly thermal plants, specialty chemicals industries, and iron ore pellets plants (IEA 2023). CastleRock Green Energy LLC, located in North America in partnership with Ensyn Corporation, generates nearly 75.7 million liters/year (US DOE 2019). Bioenergy AE Cote-Nord, a commercial pyrolysis plant company located in Canada currently generates about 38 million liters of pyrolysis oil from about 65,000 dry tons of sawmill wood waste per year via Ensyn's RTP fast pyrolysis technology (IEA 2023).

In 2011, about 1.8 million liters of pyrolysis oil was generated from about 1270 tons of plastic waste by JBI Inc. where the oil was mainly used as fuel oil for thermal plants and bioelectricity generation (Waste Management World Magazine 2013). A pyrolysis demonstration facility located in China which has been in operation since 2013, is presently using the downdraft circulating fluidized bed fast pyrolysis technology in generating about 8 million liters of pyrolysis oil per year from rice husk (Cai and Liu 2016). In India, Shell Catalysts and Technologies which was commissioned in 2018, is actively operating a hydropyrolysis facility which uses about 5 tons/day of forestry, agricultural, and other waste lignocellulosic materials to produce about 0.5 million liters of transportation fuel per year (IEA 2021).

California's fast pyrolysis demonstration plant, Biogas Energy Ltd., located on a waste management site in Sacramento, United States, is currently operational utilizing about 500 kg/h of wood waste, tree chippings, forest residues, and other agricultural waste materials in its ablative reactor modular system to generate about 65,000 liters of pyrolysis oil per year (IEA 2021).

In China, the University of Science and Technology in Hefei has developed a pyrolysis demonstration-to-commercial scale plant where about 15 tons of agricultural waste per day is transformed via fast pyrolysis into bio-oil and biochar with about 50–60% efficiency (Mulligan et al. 2015). In the same facility, the biochar is thermochemically converted into heat which is used as energy source in the plant (Mulligan et al. 2015; IEA 2021).

Globally, pyrolysis oil production capacity currently stands at about 650 million liters/year which is consumed mostly by the transportation industries, thermal plants, metals, and iron ore companies, and biochemicals industries (IEA 2021). As a result of the Renewable Fuel Standards (RFS), which is a clean energy policy in the United States mandating the blending of fossil fuel with biofuel by 2022, the United States has contributed to its biofuel market by generating about 360–372 million liters of pyrolysis oil from waste materials in 2020 (IEA 2020). Table 2.1 outlines some commercial operating companies utilizing different thermochemical conversion technologies to transform waste into value-added bioproducts.

[3] This feedstock amount is supplied by only one of Ensyn's production facilities.

Table 2.1 Pyrolysis oil production plants in the world

| Company name | Production capacity | Product | Feedstock | % Pyrolysis oil yield | Utilization of the pyrolysis oil | Location |
|---|---|---|---|---|---|---|
| Ensyn Corporation | 131.7 million liters/ year | Bio-crude (pyrolysis oil) as heating fuel | Woody biomass<br>Forest and agricultural residues (e.g., eucalyptus) | 70–75 | • Heating oil replacement in boilers<br>• Specialty chemicals | Wisconsin (United States), Brazil, Ontario, and Quebec (Canada) |
| BTG Bioliquids | 20 million liters/ year | Pyrolysis oil | Forest and agricultural residues | 30–70 | • Natural gas replacement in steam boilers | Finland, Sweden, and the Netherlands |
| Plastic2Oil/JBI Inc. | 1.8 million liters/ year<br>13 MW electricity/ year | Pyrolysis oil<br>Bioelectricity | Plastic wastes | – | • Fuel Oil<br>• Bioelectricity generation | United States |
| Agilyx Inc. | 150,000 Mt plastic waste/year | Pyrolysis oil | All types of plastic waste<br>Tires | 35–45% oil | Upgrade to low carbon oil | Indonesia, Libya, Cyprus, etc |
| Enerkem Inc. | 285 million liters/ year | Biodiesel, marine fuel, biochemicals, bioethanol | Plastic wastes<br>MSW | – | • Heating oil replacement in boilers<br>• Specialty chemicals | Wisconsin (United States), Ontario and Quebec (Canada) |
| Red Rock Biofuels | 57 million liters/ year | Jet fuel<br>Biodiesel | Woody biomass<br>Forest residues | – | • As jet fuel | North America |
| Bioenergy AE Cote-Nord | 38 million liters/ year | Pyrolysis oil | Sawmill wood waste | – | Sold to industries for heating applications | Canada |

### 2.2.2 Global Production and Consumption Profile of Other Liquid Biofuels

The main type of liquid fuel (apart from pyrolysis oil) obtained from thermochemical conversion of waste is Fischer–Tropsch (FTL) liquids (or FTL diesel and paraffinic kerosene) which are produced by catalytic conversion of syngas generated during combustion, gasification, and pyrolysis of waste materials and biomass. However, there is a wide range of other liquid biofuels that can be obtained from the thermochemical waste/biomass conversion which includes syngas-derived liquid fuels, bioethanol, biodiesel, bio-gasoline, bio-methanol, etc. These high value liquid fuels are mainly obtained via hydrothermal liquefaction, gasification, and pyrolysis for intermediate products, followed by subsequent purification or refining processes to obtain the final products which are used in automobile industries, thermal plants, electricity production, etc.

Though there are currently few commercial plants producing bio-methanol and bioethanol/bio-gasoline, about 360,000 tons of Municipal Solid Waste (MSW) is utilized as feedstock by Enerkem Inc., a Canadian biofuel company, via its gasification technology for fuel ethanol, marine fuel, biodiesel, biochemicals (e.g., ethylene, butadiene, etc.) with a total generation capacity (all its facilities in the world) of about 285 million liters/year (NOVA Chemicals 2022). This capacity is estimated to expand as Enerkem Inc. partners with Technip Energies in France, and Shell and Port in the Netherlands for improved technological solutions for efficient bioliquids production and potential $CO_2$ emissions reduction of about 165,000 t$CO_2$ per year (NOVA Chemicals 2022).

The production of renewable jet fuel and biodiesel from the gasification of woody biomass and subsequent Fischer–Tropsch and hydroprocessing technologies have been realized by Red Rock Biofuels Company located in North America, which operates at a capacity of about 57 million liters/year (US DOE 2019). Likewise, Velocys Renewable LLC,[4] a company whose headquarters is located in England, generates sustainable aviation fuel via Fischer–Tropsch process using waste materials (such as MSW, woody biomass, etc.), and it operates at a capacity of about 12 reactors each with about 52,500 square feet (IEA 2021). Bayou Fuels is a Velocys Renewable LLC's facility in the United States which produces about 36 million gallons per year of Fischer–Tropsch aviation fuel from waste biomass, and this capacity can offset greenhouse gas (GHG) emissions by about 1.3 million tons of $CO_2$ per year (IEA 2021). In 2021, Velocys reported a successful first use of its blended Fischer–Tropsch aviation fuel to power a commercial flight (IEA 2021).

In the United States, Range Fuels Inc., with a production capacity of about 379 million liters of liquid biofuels per year, was utilizing lignocellulosic waste like wood chippings, grasses, and corn stover via its two-step thermochemical conversion processes (i.e., gasification into bio-syngas and further catalytic processing into bioethanol and bio-methanol) (Renewable Energy World 2007). Unfortunately,

[4] Velocys Renewables LLC has over 90 operational facilities located across Northern America, Europe, India, and Canada (IEA 2021).

Range Fuels Inc. has not been in operation for a decade now due to the challenging issues of the biofuel market (Renewable Energy World 2007).

Arbios Chuntoh Ghuna, a biofuel company is set to start the operation of a commercial scale hydrothermal treatment to convert waste materials (such as forest residues) into bio-crude oil for transportation fuels by the end of 2024 (IEA 2021). Arbios facility located in Canada will utilize about 25,000–50,000 dry tons of woody biomass and agricultural wastes to generate about 8 million tons of HTL (hydrothermal liquefaction) bio-crude annually (IEA 2021). Another Arbios facility whose HTL technology is provided by Licella Pty Ltd., and located in Australia is set to expand its demonstration plant into a small commercial scale HTL plant of about 1.6 million liters of bio-oil annually (IEA 2021).

Metro Vancouver, a thermochemical conversion company located in Canada began the construction (in 2022) and operation (in 2024) of the world's first wastewater HTL demonstration-to-commercial plant utilizing about 2 dry tons per day of sludge from wastewater treatment plant to produce bio-crude for transportation fuel (IEA 2021). A similar demonstration plant owned by Crossbridge Energy, a former Shell refinery, is preparing to set up its HTL facility in Fredericia, Denmark to produce about 5 million liters of bio-crude from about 4000 dry tons of wastewater sludge per year (IEA 2021). Reliance Industries Ltd. in India which was constructed in 2016 is currently operating a demonstration plant which utilizes food waste, wastewater sludge, and algae biomass as feedstocks to produce about 30,000 L of HTL bio-oil for heating, transportation fuels, and power generation (IEA 2021). Another small commercial plant owned by Silva Green Fuel located in Norway is currently producing about 1.5 million liters per year of HTL bio-crude from forest residues. In the near future, Silva Green Fuel is set to expand its capacity to a commercial scale generating close to 2,000 barrels per day of bio-crude which can be used for heating and electricity generation (IEA 2021). In 2016, Altaca Energy, located in Turkey constructed and began operation of a highly efficient 8.7 million liters per year capacity of HTL bio-crude oil demonstration plant which utilizes organic waste such as food waste, sewage sludge, agricultural waste, etc. (IEA 2022). An HTL demonstration-to-commercial plant, Muradel Pty Ltd. in Australia[5] which was designed to operate on small commercial scale came to halt due to scientific and technological challenges which made scale-up impossible (ARENA 2013).

In Australia, at a production capacity of about 22,000 L of biofuel per day, Licella Holdings Ltd., formerly called Ignite Energy Resources (Victorial Coal Resources) uses catalytic HTL technology (called Cat-HTR) to convert waste (e.g., plastic waste) into high yield bio-crude which is sold to thermal plants and chemical industries for further processing into biofuels and bioenergy (Globe Newswire 2021).

[5] Muradel Pty Ltd. HTL facility was designed pre-commercialized demonstration plant in 2013 to produce HTL bio-crude oil and biochemicals from sewage sludge, algae, and recycled tires but the technology failed, and the plant became inactive in 2019 (ARENA 2013).

## 2.3 Global Outlook of Gaseous Biofuels from Waste via Thermochemical Conversion

### *2.3.1 Bio-Syngas: Global Production and Consumption Outlook*

Bio-syngas (also called bio-synthetic gas or producer gas) is a gaseous fuel generated by thermochemical waste or biomass conversion technologies such as gasification, pyrolysis, combustion, etc. Depending on the source of feedstock, bio-syngas comprises about 22 mol% carbon monoxide (CO), about 11–16 vol% hydrogen, about 5 vol% methane ($CH_4$), about 12–19 vol% $CO_2$ and about 51 vol% nitrogen as well as traces of hydrogen sulfide ($H_2S$) and water vapor with heating values ranging from about 4–5 MJ/Nm$^3$ dry (Giuliano et al. 2020; Hasler and Nussbaumer 1999). Bio-syngas has many applications in power plants and biochemicals industries where it can be purified into methane, methanol, gasoline, dimethyl ether, etc., as environmentally friendly alternatives to fossil-based natural gas (Galadima and Muraza 2015).

Currently, the share of gaseous fuels in electricity generation globally is about 10% of which bio-syngas alone contributes about 0.5% (Sikarwar et al. 2017). Bio-syngas is estimated to have a production cost of about US$15–20 per GJ which is projected to decrease in the next 10 to 15 years due to technology improvement and government subsidies (IEA-ETSAP 2013).

Several companies have developed various thermochemical conversion technologies for bio-syngas generation on large-scale production. For instance, Burkhardt Energy and Building Technology GmbH, located in Germany, provides gasification technologies for controlled high-quality bio-syngas generation from waste pellets to most energy production companies in the world (Higman 2014).

Rainbow Bee Eater, a bio-syngas company located in Australia has developed a large-scale efficient thermochemical conversion process called *Echo2* technology which converts waste such as trimmed branches of trees, forest and agricultural residues, grass chippings, construction wood waste, and cereal straw into synthetic gas (which comprises about 15 vol% of hydrogen) and sells to other companies for heat and energy generation (IEA 2022). This bio-syngas company, which started operation in 2019 with a capacity of about 5000 tons/year waste, can provide about 700 kW of hot water, 2,406 kWh electricity, 250 kg/h wood vinegar, and 200 kg/h biochar to feed nearby industries (IEA 2022). Also, in 2020, Cortus Energy AB located in Sweden started producing bio-syngas on commercial scale via gasification at a production capacity of about 5,000–15,000 tons per year (Bioenergy International 2020).

### 2.3.2 Global Demand and Supply of Biohydrogen

Thermochemical conversion of waste materials generates syngas which contains about 11–16 vol% of hydrogen gas depending on the type of waste and process conditions. The syngas can be purified to isolate pure hydrogen gas using technologies such as water shift reaction, swing adsorption, solvent scrubbing, electrolysis, etc. (see Chap. 3). Hydrogen gas has four different 'shades' based on how they are produced, namely:

- *Brown hydrogen* is generated from coal without emission control.
- *Grey hydrogen* is produced from natural gas without emission control.
- *Blue hydrogen* is generated from natural gas with emissions captured.
- *Green hydrogen* or *biohydrogen* is obtained from biomass waste or other renewable energy sources via electrolysis.

Today, almost all the hydrogen consumed in the world is grey hydrogen at a consumption rate of about 90 million tons per year (Global Energy Perspective 2023). However, as the demand for clean hydrogen increases over the years, grey hydrogen demand would decline. It is projected that by 2050, the demand for green hydrogen would increase to about 125–585 million tons per year (Global Energy Perspective 2023).

Hydrogen gas is used in fuel cells to generate electricity, as transportation fuel, as raw material to produce synthetic fuels as well as additives to reduce the sulfur content of fuels. In 2021, Atmosfair company located in Emsland, Germany, introduced the first of its kind, 25,000 L per year of synthetic aviation fuel and green kerosene production facility which utilizes biohydrogen from thermochemical conversion processes and other sources of renewable energy (IEA 2022). According to the projections made by the International Energy Administration in 2022, the world could see an estimated global annual production of about 2 billion synthetic kerosene by 2030 which could replace fossil-based fuels in the aviation industry reducing about 0.5–1.5% of global demand of fossil fuels (IEA 2022).

Despite technological challenges and uncertainties in commercializing bio-based hydrogen gas for fuel, the global eco-friendly hydrogen demand is projected to rise significantly by 2050. Presently, the global annual demand for hydrogen gas is about 90–94 million tons which forms about 2.5% of the world's total primary energy consumption (IEA 2022). Global hydrogen demand is projected to reach about 115 million tons by 2030 which would be on the right track for net zero emissions by 2050 (IEA 2022). Hydrogen gas produced from fossil-based natural gas forms nearly 98.8% of the world's total demand annually (IEA 2022). It is projected that by 2050, biohydrogen demand could account for about 73–99% (125–585 million tons per year) of total hydrogen demand in the world, as the world gradually embraces renewable energy resources for sustainability (IEA 2022). The projected increase in biohydrogen demand in the world could displace about 10 billion cubic meters of natural gas demand (IEA 2022).

According to the International Energy Administration, the world's hydrogen demand saw about 5% increase in 2021 (IEA 2022) where most of the demand was for various applications in the chemical industries. In 2021, the applications of hydrogen in the transport sector, power generation, and building and Construction industries were low at 40,000 tons which is about 0.04% of global hydrogen demand (IEA 2022).

China is presently the leading consumer of hydrogen with about 28 million tons consumed per year since 2020 (IEA 2022). The United States and the Middle East are the next major consumers of hydrogen in the world taking a share of about 12 million tons per year (IEA 2022).

There are several commercial biohydrogen production companies in the world that utilize waste materials like plastics, agricultural wastes, etc., as feedstock. For instance, Verennes Carbon Recycling facility in Canada, with a capacity of 125 million liters per year, is set to start operation in 2025 and will utilize biofuels generated by thermochemical biomass waste conversion for biohydrogen generation via 88 MW electrolyzers (IEA 2022). Similar demonstration and commercial biohydrogen plants are underway to full operation in France (Total Energies SE having about 125 MW electrolyzers) and Spain (Repsol SA with 250,000 tons of biofuel to be purified into biohydrogen) (IEA 2022).

## 2.4 Global Outlook of Solid Biofuels from Waste via Thermochemical Conversion

Thermochemical conversion technologies like gasification, hydrothermal carbonization, pyrolysis, combustion, carbonization, and torrefaction mostly generate solid by-products like biochar, hydrochar, charcoal, etc., which can be used as sources of fuels '*as are*' or processed further into fuels like pellets, briquettes, etc. The different kinds of solid biofuels have been discussed in detail elsewhere (Ofori-Boateng 2022). Figure 2.2 shows pictorial images of pellets, briquettes, and biochar as solid biofuels.

**Fig. 2.2** Pictures of charcoal (**a**), Biochar/Char (**b**), and Pelleted/Briquetted Biochar (**c**) (*Courtesy* Wikipedia)

Wood waste and lignocellulosic pellets and briquettes are among the fast growing biofuel resources globally with a generation capacity of about 38 million tons in 2019 and annual generation increase of about 17%, and Europe is currently taking the largest share of about 55% (Baqir et al. 2019).

### *2.4.1 Charcoal and Charcoal Briquettes: Global Production and Consumption Statistics*

Charcoal is a lightweight and porous black carbon-rich fuel with a high heating value (HHV) between that of wood (16–20 MJ/kg) and coal (24–30 MJ/kg) ranging from about 17–33 MJ/kg depending on the type of biomass used and conditions of carbonization (e.g., temperature) (Nabukalu and Reto 2019). Charcoal is char usually produced from woody biomass, agricultural and forest residues via carbonization or slow pyrolysis which results in GHG emissions, particulate matter, nitrous oxide, carbon monoxide, etc. (Pennise et al. 2001). One kilogram of charcoal-produced unsustainably is found to contribute about 6–9 kg $CO_{2eq}$ over its life cycle (FAO 2023). Charcoal is the most abundantly utilized fuel source for cooking and heating in most parts of the world (especially the developing countries) and contributes about 14–16% of total global energy use annually (Baqir et al. 2019). In Africa for instance, charcoal is produced mostly by smallholders by means of a conventional clay furnace (Mensah et al. 2022).

Charcoal briquettes are produced by blending powdered charcoal with binders such as clay soil to enhance their heating values. Pellets and briquettes made from charcoal can be used as coal and wood replacements for the generation of heat and electricity at high efficiency. Charcoal briquettes and pellets are used as sustainable and clean solid fuels for grilling and as feedstock in furnaces for power generation.

In rural Africa and most tropical regions of the world, biomass (especially charcoal) covers a significantly large share of primary energy consumption with electricity contributing about 2–4% of total energy mix (Baqir et al. 2019). The demand for charcoal in most rural areas of the world as a cooking fuel and heat generation source has resulted from the limited supply of fossil-based energy. However, the demand for charcoal currently has increased on the global market not because of consumption in rural areas but also in cities with electricity generation where charcoal is more convenient as fuel for grilling, etc.

In 2019, approximately 54 million tons (Mt) of charcoal made from wood, forest, and agricultural residues was generated globally with about 0.5–5 Mt increment annually (Nabukalu and Reto 2019; FAO 2023). Africa has been the major producer of charcoal globally, contributing about 66.9% (Nabukalu and Reto 2019; FAO 2023). With a total production capacity of about 15.8 Mt/year, Nigeria, Ghana, Ethiopia, Egypt, and Tanzania have been the top producers of charcoal in Africa (Oppong Mensah et al. 2024; Nabukalu and Reto 2019). However, Brazil solely produces about 11% of the world's wood charcoal annually which is basically transformed into pig iron and exported to other parts of the world (Nabukalu and Reto 2019; Njenga et al. 2023). In 2021, China's total charcoal production increased to about 99.5 million tons, a growth of about 1.7% in one month (from April to May) mainly for applications in the steel industry (Virtue Market Research 2023). The European Union imports about 1.22 million tons of charcoal from Africa and sub-Saharan Africa to add to their 0.64 million tons of charcoal generated internally per year (FAO 2023). With the production of coal set to decline soon, charcoal from waste can excellently replace coal when the production process is made more efficient.

There are many commercial production plants that utilize waste materials, especially forest and agricultural residues via thermochemical conversion processes to produce charcoal and pellets in a more sustainable way. For instance, Pinnacle Renewable Energy Inc. has over 17 pellet plants scattered in Canada and the United States which together produce about 4.9 million tons of pellets annually (Smith 2021). Also, the second largest pellet company in the world, Woodville Pellets, owned by Graanul Biotech Inc. and located in Texas in the United States generates about 496,000 tons of pellets per year for diverse fuel and energy applications. Other bioenergy companies owned by Graanul Invest Group together are currently producing about 2.98 million tons of pellets per year from forest residues and sawmill wood waste (Bioenergy International 2023). Presently, Kingsford Inc., owned by The Clorox Company, is the leading producer of charcoal briquettes in the United States holding about 80% share of the total briquettes generated globally (FAO 2023).

The world's largest producer of forest residue/wood pellets for bioenergy is Enviva Pellets LLC which has numerous facilities located in many states in the United States which together produce about 6.2–13 million tons of wood pellets per year (FAO 2023).

Though there has been a decline in the production of charcoal over the years, the total annual generation of charcoal keeps increasing consistently. Figure 2.3 shows the production trend of global charcoal production from waste from 2000 to 2022.

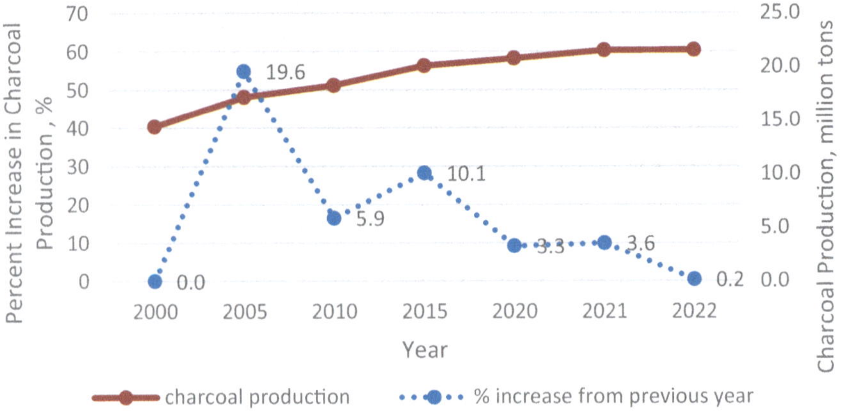


**Fig. 2.3** Global charcoal production trend from 2000 to 2022 (Statista 2024)

### 2.4.2 *Biochar, Biochar Pellets, and Briquettes: Global Statistics*

Biochar[6] is a type of charcoal which is environmentally friendly alternative to fossil-based coal used in thermal plants for bioelectricity and heat generation though some studies have reported certain challenges with biochar use in combined heat and power (CHP) plants due to their low mechanical strength and fine texture (Mohammadi 2021). The energy density of biochar (about 27–32 MJ/kg) is reported to be higher than certain types of coal (24–30 MJ/kg) and depends largely on the type of waste material and conditions of thermal decomposition (Kosakowski et al. 2020).

Biochar has numerous applications in agriculture, bioenergy, and chemical industries which can intensify carbon sequestration[7] hence reducing GHG emissions in the atmosphere. In order to improve its texture and strength, biochar has been utilized as pellets, briquettes, and electrodes for microbial fuel cells for sustainable bioenergy production (Lee et al. 2017). Biochar is utilized as an environmentally friendly fertilizer in farms to improve soil fertility and mitigate climate change.

Pelleted biochar and briquettes are energy dense solid biofuels obtained by mechanical compression of biochar into uniformly sized fuel with or without binders (Zhu et al. 2019). Biochar pellets are found to be more durable, environmentally benign, energy dense, mechanically strong, and easily combustible into bioenergy or biofuel compared to biomass pellets (Li et al. 2015; Zhu et al. 2019).

Haycarb PLC Pvt Ltd., a company whose operation facilities are located in Sri Lanka, Indonesia, and Thailand is the leading producer of coconut biochar used as activated carbon with an annual production capacity of 50,000 Mt (Haycarb Activated Carbon Solutions 2024). Founded in 2007, a biochar company, Carbon Gold Ltd., located in the United Kingdom (UK) has been using tree trimmings, cocoa farm residues, and other agricultural residues in their mobile kilns to produce about 150 kg biochar per batch from about 600 kg of the waste (International Biochar Initiative 2022).

In India, Mash Makes A/S, a commercial pyrolysis plant which started operation in 2023, is actively generating biochar from about 3000 tons agricultural residues per year (IEA 2021). Pacific Biochar Inc., located in the United States uses fast pyrolysis technology to convert forest residues from California's fire hazard areas into biochar which can remove about 2.5–2.9 tons of $CO_{2eq}$. from the atmosphere per ton biochar produced (Thengane et al. 2021). The biochar from this facility currently has many commercial applications such as compost additives, soil amendment and remediation agents.

---

[6] Depending on the type of thermochemical conversion technology used, biochar or hydrochar can be produced. Hydrochar is generated by hydrothermal carbonization whilst biochar is generated by combustion, gasification, and pyrolysis.

[7] Carbon sequestration is the capturing and storing of $CO_2$ from the atmosphere with the aim of reducing greenhouse gases hence reducing global warming and climate change.

### 2.4.3 *Torrefied Biomass/Waste: Global Production and Consumption Outlook*

Torrefied biomass/waste is highly efficient and energy dense solid biofuels whose physicochemical characteristics have been improved by torrefaction, a thermochemical conversion technology. Torrefied biomass or waste materials can further be combusted, gasified, pyrolyzed, or thermally liquefied into various kinds of liquid and gas biofuels and bioenergy (see Chap. 3 for more details on torrefaction).

In 2011, Takachar LLC was established to help farmers in the Boston area, the United States to avoid the use of open incineration of their agricultural residues, thus contributing about 98% reduction in smoke emissions (Breakthrough Energy 2023). Takachar uses small-scale torrefaction technology to convert forest residues into solid fuels, fertilizers, and other biochemicals at varying capacities depending on the torrefaction conditions.

## 2.5 Global Profile of Biochemicals from Waste via Thermochemical Conversion

Thermochemical waste conversion technologies such as pyrolysis and gasification have been used successfully to produce biochemicals such as p-xylene, toluene, benzene, etc. These biochemicals are mostly sourced from the solid by-products (e.g., biochar) of gasification, pyrolysis, hydrothermal carbonization, etc. Biochar and hydrochar are also used in agricultural industries as eco-friendly alternatives to synthetic fertilizers for soil improvement (Cha et al. 2016). In chemical industries, it is used as contaminants adsorbent, activated carbon, construction material, a precursor for biocatalysts and nanotubes production, etc. (Qian et al. 2015; Cha et al. 2016). In the metallurgy industries, biochar is utilized as a substitute for coke during the coking process as it reduces iron ore during steel production. Again, biochar is sometimes used as cost effective and green additives in metalized pellets which are used in furnaces (Khanna et al. 2019; Ye et al. 2019).

Bio-syngas can be converted into other useful products like liquid fuels, wax, and biochemicals via the Fischer–Tropsch Synthesis (FTS) process. FTS can also generate valuable chemicals like olefins, aromatics, wax, etc., via the thermochemical conversion process where bio-syngas or bioliquids are produced as intermediate products. FTS biochemicals can also be used in the production of plastics, pharmaceuticals, dimethyl ether, paint, personal care products, additives in polymer production, lubricants and greases, candles, textiles, fishing lines, etc. In South Africa, for instance, Sasol Ltd. is utilizing sustainable feedstock to produce FTS waxes and other FTS products at a capacity of about 170,000 barrels per day (Sasol 2023).

Circa Group AS, a biofuel company located in Norway has developed a 0.8 million liter per year commercial pyrolysis plant projected to start operation by the end of 2024 (IEA 2021). This demonstration facility was designed and developed to expand its capacity to about 80,000 tons of lignocellulosic waste and would purify pyrolysis oil into pure cyrene[8] (IEA 2021). Using the same Furacell technology, the pulp and paper company, Norske Skog Ltd, located in Australia is currently in partnership with Circa Group AS, and operating a 50 ton/year capacity pyrolysis facility which generates about 40,000 L per year of 99% pure cyrene from lignocellulosic waste (IEA 2021).

Benzene, toluene, ethylbenzene, and xylene (commonly called BTEX) are important volatile organic solvents commonly used as additives in petroleum industries. With environmental benefits in mind, Anellotech Inc.'s fast pyrolysis facility in Texas developed a demonstration-to-commercial plant which now uses at least 500 tons of corn stover, bagasse, and wood waste per day to produce eco-friendly BTEX[9] mixture solvent, polymers, and fuels (IEA 2021). The 90% pure BTEX as well as the green fuel produced by Anellotech Inc. can reduce greenhouse gas potentials by about 70–90% compared to petroleum-based products (IEA 2021). Again, with a total production capacity of about 400,000 tons/year, Anellotech Inc. in partnership with Axens Inc. (located in France) utilizes woody biomass and plastic waste via a pyrolysis process to produce 100% bio-marine fuels and bio-based aromatic chemicals which are used in the manufacturing of bioplastics such as polyester, nylon, etc. (IEA 2021).

## 2.6 Global Statistics of Bio-Heat and Bioelectricity Generation from Waste

Bioenergy (i.e., bioelectricity and heat) generation has derived their sources from all types of biofuels such as pellets, pyrolysis oil, etc. In 2022, the world's bioenergy capacity was about 149 gigawatts with China holding the largest share of about 23% followed by Brazil (17.2 gigawatts bioenergy) (World Bioenergy Association 2020). Biopower has seen an increasing trend from about 162 terawatt-hours in 2000 to about 685 terawatt-hours in 2020 (World Bioenergy Association 2020). In 2018, about 420 terawatt-hour (TWh) of the total global bioelectricity (i.e., 637 TWh) was generated from MSW and Industrial waste via thermochemical conversion technologies (World Bioenergy Association 2020). Comparing the 2018 production capacities of biomass-based bioelectricity via thermochemical conversion to bioelectricity from other renewable resources like solar, wind, etc., biomass bioelectricity formed only about 9.2% of the total bioelectricity mix, which is far lower than wind bioelectricity (18.5% of total renewable electricity mix) but higher than geothermal (1.3% of total renewable electricity mix) and solar bioelectricity (8.2% of total renewable electricity mix) (Global Bioenergy Statistics 2020). Currently, Europe is the world's leading producer of heat and bioelectricity from biomass and other waste materials

[8] Cyrene is a colorless viscous eco-friendly solvent used in pharmaceutical industries.

[9] The BTEX solvent mixture can be purified to extract p-xylene to produce bio-PET (polyethylene terephthalate) which has been successfully demonstrated by Suntory, Annelotech's partner company.

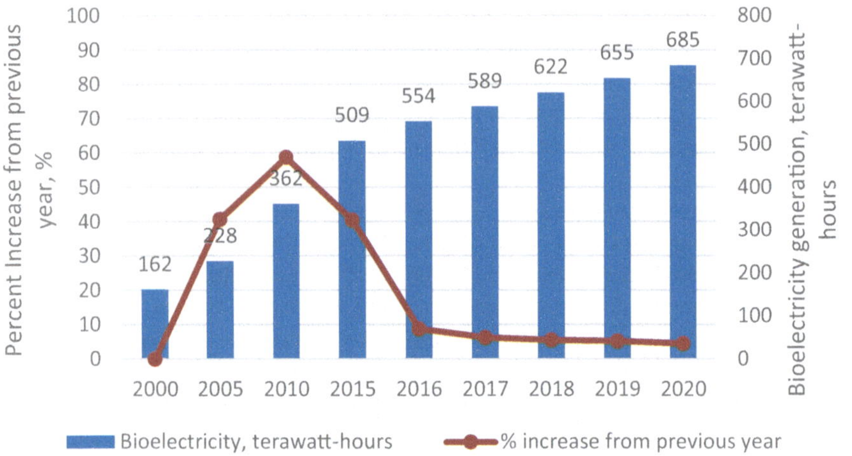


**Fig. 2.4** Global bioelectricity generation from 2000 to 2020

contributing a share of about 87% of the total global bio-heat generation of about 1.12 exajoules (EJ) per year (Global Bioenergy Statistics 2020). Figure 2.4 shows the percent increase trend in biomass bioelectricity generation from 2000 to 2020.

Combined heat and power (CHP) plants or cogeneration plants are developed to transform waste materials into heat and electricity. Europe has consistently dominated the biomass-to-heat generation by CHP plants accounting for about 82% of the global heat generation followed by the United States taking a share of about 44% in 2018 (World Bioenergy Association 2020). Biohydrogen plays a negligible role as a source of fuel for bioelectricity generation as it accounts for only about 0.2% of the total global electricity generation (IEA 2022). Table 2.2 summarizes the production capacities of MSW incineration and combustion in the United States.

Incineration systems have also contributed to bioelectricity and heat generation capacities in the world. There are about 1249–1300 incineration facilities for MSW treatment in the world with about 93% located in Japan (Cucchiella et al. 2017). About 77–80% of the total incineration systems in the world generate electricity (50 59 facilities) whilst the rest are used as CHP plants (Cucchiella et al. 2017). The United States currently operates about 35 gasification facilities utilizing mainly fossil-based fuel and negligible amount of biomass or waste (Mukherjee et al. 2020).

Table 2.2 Current incineration and combustion technologies employed in the United States

| Company | Location | Amount of waste (ton/year) | Electricity generation (MWh) | Number of people it serves |
|---|---|---|---|---|
| Engen LLC[10] | Florida | 182,500 | 32,850 | 169,560 |
| City of Spokane Waste-to-Energy facility[11] | Washington DC | 292,000 | 56,940 | 426,347 |
| Ecomaine[12] | Maine | 200,750 | 32,193 | 250,000 |
| Mid-Maine Action Corporation | Maine | 91,250 | 10,950 | 65,000 |
| ESOCO Orington Inc. or PERC Holdings LLC | Maine | 262,800 | 25 MW | 400,000 |
| Great River Energy | Minnesota | 365,000 | 54,750 | 850,000 |
| GRE HERC Services and others | Minnesota | 73,000–442,380 | 1095–80,373 | 42,000–1,156,212 |
| Covanta Holding Corporation[13] | Many locations | 21 million | 9 million | – |
| Wheelabrator Technologies[14] | Many locations | 7.4 million | 5.8 million | – |

Adopted from Mukherjee et al. (2020) and Michaels and Krishnan (2019)

## 2.7 World's Market of Biofuels and Bioenergy from Thermochemical Conversion

### *2.7.1 Global Pyrolysis Oil Market, Share, and Growth*

The world's pyrolysis oil market is projected to expand consistently over the coming years owing to the high call for eco-friendly and sustainable energy resources to reduce the current $CO_2$ emission burdens from fossil fuels on the environment. Currently, about 60% of the total pyrolysis oil market is dominated by the fuel sector (Global Market Insight 2023).

With a steady annual pyrolysis oil market worldwide growing from US$ 3.1 billion in 2021 to US$ 3.2 billion in 2022, the world is set to achieve a compound annual

[10] Mukherjee et al. (2020).

[11] Mukherjee et al. (2020).

[12] Mukherjee et al. (2020).

[13] Themelis and Mussche (2013).

[14] This facility generates about 1123 tons of $NO_x$ per year and uses about 1093 °C combustion temperature (Owens and Szczepkowski 2010).

growth rate (CAGR[15]) of 4.3% by 2030 (Global Market Insight 2023). Industries like thermal plants, transportation, pharmaceuticals, and biochemical industries have always accounted for a large portion of the global pyrolysis oil market. Plastic waste pyrolysis oil contributed the largest share of the total pyrolysis market (US$ 641 million at CAGR of 5.3%) followed by tire bio-oil due to the growing interest in reusing and recycling plastic and tire wastes for environmental sustainability (Global Market Insight 2023). Currently, the dominant feedstocks for pyrolysis production on commercial scale in the world are plastics, rubber, tire, wood, oil sludge, and other lignocellulosic waste materials.

The tire pyrolysis oil market size stood at about US$ 329 million in 2022, and it is projected to increase steadily at a CAGR growth of 5.4% until 2030 (Global Market Insight 2023). Continuous improvement of pyrolysis technology over time would affect product quality and marketability positively. The prominent company drivers of tire pyrolysis oil market in the world presently include Alterra Energy Inc., Ensyn Corporation, Bioenergy AE Cote-Nord, Bridgestone Corporation, Green Fuel Nordic Oy, Pyrocell, Shenyang Dongxing Chemical Industry Co. Ltd., Quantafuel ASA, New Energy kft, New Hope Energy, Nexus Circular, Trident Fuels Pty Ltd. in South Africa, and Mk Aromatics Ltd. (IEA 2022; Market Research 2022).

Considering the application and regional contributions of bioenergy production to the global pyrolysis oil market, Asia Pacific takes the greatest shares. Factors that facilitate the growth of the world's pyrolysis oil market include:

- Renewable energy policies and subsidies from governments,
- Advanced technology and infrastructure,
- Continuous improvement through research and development,
- Global attention on waste management,
- Rising concerns about environmental pollution and climate change from the use of fossil-based crude oil,
- Increase in circular economy investments,
- Fast industrialization,
- Strategic collaborations among stakeholders in industries, universities, and government bodies, and
- Increase in world's population growth.

According to the International Energy Administration, the top 20 company drivers of pyrolysis oil market growth in the world are Ensyn Corporation in Canada, Anellotech Inc. in United States, Beston Group Co. Ltd. in China, BTG-BTL in Netherlands, Chemelot in the Netherlands, Plastic Advanced Recycling Corporation in United States, Pryme in Netherlands, BP plc in United Kingdom, TotalEnergies in France, Pyro-oil Nigeria Ltd., Cepsa in Spain, Valero Energy Corporation in United States, Exxon Mobil Corporation in United States, Chevron Corporation in United States, Suez SA in France, Royal Dutch Shell plc in Netherlands and United

[15] CAGR is a financial index to measure the average annual growth rate of an investment or any valuable unit over a specified period with the assumption that all profits would be reinvested. CAGR is usually used to compare returns for anything that can increase or decrease in value over time.

Kingdom, BASF SE in Germany, Repsol SA in Span, Pyrowave Inc. in Canada, Plastic Energy Ltd. in United Kingdom, Renewable Energy Group Inc. in United States, Agilyx Corporation in United States, Neste Corporation in Finland and BTG Bioliquids BV in Netherlands, with the United States and Canada having the largest share due to advanced technology, implementation of sustainability initiatives and efficient government regulations (IEA 2021; Market Research 2022). The market of pyrolysis oil from waste may fluctuate due to some major challenges such as:

- High cost of pyrolysis oil production.
- Competition from other biofuel sources.
- Lack of consumer awareness which may result in demand decline.
- Government and regulatory barriers.

Intensification and implementation of strategies which could potentially increase the market share of pyrolysis oil for biofuels may change the fluctuations or probably drastic decline in the global market for thermochemical conversion bioproducts in general.

### 2.7.2 *Hydrothermal Treatment Oil: Global Market Statistics*

The hydrothermal treatment of waste (comprising hydrothermal liquefaction, hydrothermal carbonization, and hydrothermal gasification) into various products including biofuels and biochemicals has seen sharp expansion in its market globally since 2022, and it is projected to reach US$ 1.1434 billion by 2029 (from about US$ 550 million in 2022) at a CAGR of 12.4% which is very high compared to pyrolysis oil market (IEA 2022). Presently, hydrothermal liquefaction (HTL), hydrothermal carbonization (HTC), hydrothermal gasification (HTG), and hydrothermal refining (HTR) are the top technologies that dominate the global hydrothermal treatment market. North America, Asia, and Europe are currently the leading investors in commercial and pilot scale hydrothermal biomass treatment market with key company players including Genifuel Corporation in the United States, TerraNova Energy GmbH in Germany, and Veolia Energy in Finland.

### 2.7.3 *Global Biomass Gasification Market and Growth Potentials*

The conversion of waste via gasification technology generates products such as biosyngas, biohydrogen, biochemicals, biochar, etc. Coal and fossil-based fuels have dominated the gasification market since time immemorial, but the world has seen a gradual decline due to the emergence of biomass feedstock being converted into biofuels via gasification.

The world's biomass gasification market is anticipated to grow as the world gradually transitions into renewable energy sources due to environmental concerns with the use of fossil fuels. The global gasification market was estimated at US $96 billion in 2022 and with a CAGR growth of 7.2–8.9%, it is projected to reach US $194.5 billion in 2027 (Markets and Markets 2022; Global Insight Services 2023). Between 2019 and 2023, the global biomass gasification market was maintained steadily at a CAGR of 11.1% during which the world recorded a consistent rise in renewable energy demand (Global Insight Services 2023). The gasification market is set to rise exponentially due to the high generation of waste that is readily available daily because of the increased world population. Again, with the implementation of government incentives in promoting environmentally friendly fuels especially in Europe and North America where the focus is so strong on displacing fossil fuel, gasification technologies to convert waste to syngas would witness many investors. However, the high capital and operation costs of gasification, fluctuations in global energy costs as well as technological challenges (such as tar formation during syngas production) may hinder the linear growth pattern of gasification market worldwide. Research and development to improve the efficiency and scalability of gasification technology would help curb the hurdles of technology failure, especially during scale-up.

Considering the type of waste used as the feedstock for gasification, forest and agricultural residues are projected to have the largest market share in the near future due to the increasing number of plants that utilize forest residues (because of their high cellulose and organic matter content) in combined heat and power plants for bioenergy production. The major application of syngas and biochar generated during gasification is transportation fuels such as biodiesel and bio-gasoline. The key company drivers in the world's bio-syngas via gasification market are Enviva Enterprise which converts wood pellets to syngas, Goteborg Energi AB in Sweden, Ankur Scientific Energy Technologies Pvt Ltd. in India, Valmet Corporation in Finland, EQTEC plc in Ireland, and KASAG Swiss AG in Switzerland.

Like all biofuels, the global green hydrogen market is driven by technological advancement, sustainability, and government initiatives. As the world is gradually gravitating toward energy security, sustainable development, carbon emissions reduction, and cost effective biofuels, biohydrogen demand and market are projected to progress over the years with decision makers and investors staying ahead of the competitive curve.

### *2.7.4 Global Fischer–Tropsch Market and Growth Potentials*

FTS wax has many applications in the chemical manufacturing industries where they are used as additives for waterproof jackets, tents, paints, and seals. The global FTS wax market size was valued at US$ 790 million in 2021 at a CAGR of 6.7% and is projected to reach US$ 1.416 billion by 2030 (Spherical Insights LLP 2022; Market Research 2024) which is higher than HT and pyrolysis oil markets due to the vast diverse applications of wax. In the United States, the FTS wax market stood at US$ 245 million in 2021 which was about 31% of the world's market share (Spherical Insights LLP 2022). The United Kingdom and France are among the major consumers of FTS wax in the world, and these countries are presently contributing about 19% and 16% of the total FTS wax market, respectively (Spherical Insights LLP 2022). The rapid market growth of FTS wax is due to the increase in demand for bio-based and inexpensive chemicals for wax applications, paints, and seals. This market can grow further through advancement in FTS technology through innovative research and development (R&D) as well as increased demand for biochemicals which would propel massive investment into the market. The potential for using FTS wax as additives in tire production to reduce cracks and aging process is very high especially when eco-friendly products are used. Thus, the FTS wax market target for tire processing is projected to grow at a rate of 12–13% from now till 2030 as the automobile industry expands (Spherical Insights LLP 2022). The major leading company drivers in FTS wax via R&D, demonstration, and commercial scales development include Sasol South Africa, AWAX Group in Europe, Evonik Industries AG, BASF SE, and DEUREX AG in Germany, Nippon Seiro Co. Ltd. in Japan, Honeywell International Inc. and Exxon Mobil Corporation in the United States, Petrobras in Brazil, and Royal Dutch Shell PLC in the Netherlands.

The global market for FTS paraffinic kerosene is growing rapidly due to the high demand for green aviation fuel to meet carbon reduction goals. The world's FTS kerosene market is expected to reach US$ 4.637 billion by 2030 at a CAGR of 11% compared to the market size in 2023 which was US$ 2.475 billion (Research Nester 2024; Market Research 2024). In 2023, Europe, Asia Pacific, and North America held the greatest share of FTS kerosene market at values of US$ 853.8 million, US$ 853.8 million, and US$ 556.6 million, respectively, due to the stringent implementation of bioenergy regulations in these regions (Market Research 2024). The predominant company drivers in the world's FTS kerosene market are Sasol Ltd. (in South Africa, the United States, etc.), Fulcrum Bioenergy,[16] Chevron Corporation, Petro SA, and Shell Global, and these companies together contribute about 84.6% of the total FT kerosene or sustainable aviation fuel market share (Market Research 2024).

### 2.7.5 *Global Charcoal and Biochar Market and Growth Potentials*

The consumption of charcoal as a source of fuel in most regions of the world depends on many factors such as affordability, cultural practices, cost of living, etc. Based on reports by the Food and Agriculture Organization (FAO), in 2020, the world's charcoal imports totaled about US $95 billion with Asia and Europe accounting

[16] Fulcrum Bioenergy is a green energy company located in the United States which converts landfill wastes gases and garbage of all kinds into FT synthetic paraffinic kerosene/jet fuel at a production capacity of about 400 million gallon per year (Market Research 2024).

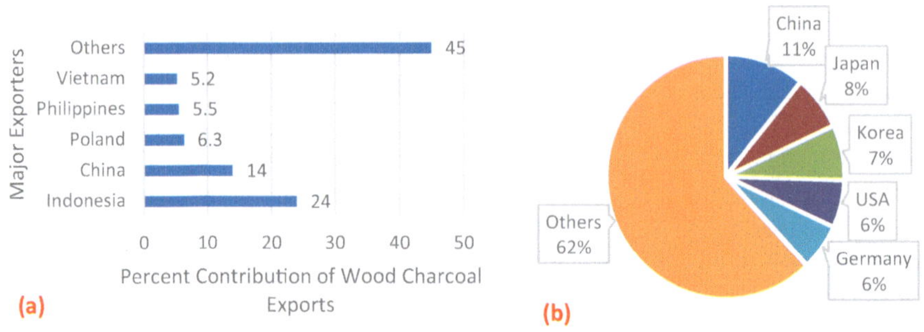


**Fig. 2.5** Global Exports (**a**) and Imports (**b**) of Wood Charcoal in 2022 (FAO 2023)

for about 95% (FAO 2023). In 2022, the estimated average wood charcoal imports for about 130 countries in the world was about $1.7 billion, about 11% increase compared to the previous year (Trend Economy 2024). China has been one of the countries that contributes largely to the market shares of both exports and imports of wood charcoal in the world. Most countries located in Europe and Asia are the major contributors to wood charcoal exports and imports market share globally. The United States imported about US$ 165 million worth of charcoal from Brazil, Namibia, and Argentina in 2020 during the summer season when barbecue is predominant (Trend Economy 2024). Figure 2.5 shows the key contributors of wood charcoal imports and exports in the world.

A recent study on the value of biochar and wood vinegar reports that the average price for biochar as quoted by users was $1,807 per ton, with prices ranging from $100 to $6,750 per ton (Robb et al. 2020). This depends on the grade of biochar, moisture content, and end-use. The global charcoal market is projected to reach US$ 11.1 billion at a CAGR of 4.1% by 2030 with an estimated increase of about 32% from 2024 (Virtue Market Research 2023).

## 2.8 Conclusion

Thermochemical conversion technologies such as pyrolysis, hydrothermal liquefaction, gasification, etc., have been employed successfully on both demonstration and commercial scale levels for the conversion of waste materials such as tires, plastic waste, and lignocellulosic biomass at high yields. The growth margin of biofuels, bioenergy, and biochemicals worldwide is increasing as their market also soars from time to time. Despite the market expansion, advancement, and improvement in biofuels, there are still obstacles impacting the overall production, consumption, and market opportunities in the world hence fighting off probable investors and consumers. This chapter reviewed the world's generation and consumption statistics of biofuels and biochemicals generated via thermochemical conversion technologies.

Also, the major market drivers, market prospective, potential financiers, topographical distribution, and growth forecast of biofuels, bioenergy, and biochemicals generated from waste via thermochemical conversion processes were elaborated in this chapter.

## References


Australian Renewable Energy Agency (ARENA) (2013). Bioenergy/energy from waste: taking crude to the next level. https://arena.gov.au/projects/advancing-marine-microalgae-biofuels-to-commercialisation/. Accessed 20 Dec 2023

Awogbemi O, Kallon DVV (2022) Valorization of agricultural wastes for biofuel applications. Heliyon 8(10):e11117

Baqir M, Kothari R, Singh RP (2019) Fuel wood consumption, and its influence on forest biomass carbon stock and emission of carbon dioxide: a case study of Kahinaur, district Mau, Uttar Pradesh, India. Biofuels 10:145–154

Bioenergy International (2020) Cortus Energy produces first biosyngas in Hoganas. https://bioenergyinternational.com/cortus-energy-produces-first-biosyngas-in-hoganas/. Accessed 11 Mar 2024

Bioenergy International (2023) Graanul Invest loads its largest US pellet shipment. https://bioenergyinternational.com/graanul-invest-loads-its-largest-us-pellet-shipment/. Accessed 19 Mar 2024

Breakthrough Energy (2023) Takachar is on a mission to transform agriculture globally. https://breakthroughenergy.org/news/takachar/. Accessed 12 Mar 2024

Cai W, Liu R (2016) Performance of a commercial-scale biomass fast pyrolysis plant for bio-oil production. Fuel 182:677–686

Cha JS, Park SH, Jung SC, Ryu C, Jeon JK, Shin MC et al (2016) Production and utilization of biochar: a review. J Ind Eng Chem 40:1–15

Chen X, Onal H (2016) Renewable energy policies and competition for biomass: implications for land use, food prices, and processing industry. Energy Policy 92:270–278

Cucchiella F, D'Adamo I, Gastaldi M (2017) Sustainable waste management: waste to energy plant as an alternative to landfill. Energy Convers Manag 131:18–31

Doshi A, Pascoe S, Coglan L, Rainey TJ (2016) Economic and policy issues in the production of algae-based biofuels: a review. Renew Sustain Energy Rev 64:329–337

Emery E, Sanscartier D, Jansen R, Saskatchewan (2021) Assessing the use of liquid biofuels in the Northwest territories. Saskatchewan Research Council (SRC) Final Report. https://www.inf.gov.nt.ca/sites/inf/files/resources/src_nwt_biofuels_final_report.pdf. Accessed 18 Aug 2024

Food and Agriculture Organization (FAO) (2023) FAOSTAT statistical database. Food and Agriculture Organization of the United Nations, Rome. https://www.fao.org/faostat/en/#data/FO. Accessed 12 Oct 2023

Galadima A, Muraza O (2015) From synthesis gas production to methanol synthesis and potential upgrade to gasoline range hydrocarbons: A review. J. Nat. Gas Sci. Eng. 25:303–316.

Giuliano A, Freda C, Catizzone E (2020) Techno-economic assessment of bio-syngas production for methanol synthesis: a focus on the water-gas shift and carbon capture sections. J Bioeng 7(3):70

Global Energy Perspective (2023) Hydrogen outlook. https://www.mckinse.com/industries/oil-and-gas/our-insights/global-energy-perspective-2023-hydrogen-outlook. Accessed 29 Jan 2024

Global Insight Services (2023) Biomass gasification market analysis and forecast to 2032: by application (chemicals, liquid fuels, power, gas fuels), gasifier type (moving or fixed bed, fluidized bed, entrained, and others), source (solid biomass, liquid biomass, biogas, municipal waste, others), and region. https://www.globalinsightservices.com/reports/biomass-gasification-market/#:~:text=Some%20major%20key%20players%20of%20Biomass%20Gasification%20Market,Valmet%20Corporation%20%28Finland%29%2C%20and%20Vaskiluoto%20Voima%20Oy%20%28Finland%29. Accessed 6 Jan 2024

Global Market Insight (2023). Plastic waste pyrolysis oil market size by feedstock, by process, by end use and forecast 2024–2032. https://www.gminsights.com/industry-analysis/plastic-waste-pyrolysis-oil-market

Globe Newswire (2021) Licella Cat-HTR technology. https://www.globenewswire.com/en/news-release/2021/09/14/2296564/0/en/Global-Technology-Pioneers-Licella-Announce-Commissioning-of-World-s-Largest-Operating-Hydrothermal-Liquefaction-Facility.html. Accessed 4 Nov 2023

Hasler P, Nussbaumer T (1999) Gas cleaning for IC engine applications from fixed bed biomass gasification. Biomass Bioenergy 16:385–395.

Haycarb Activated Carbon Solutions (2024). https://www.haycarb.com/. Accessed 13 Jan 2024

Higman C (2014) State of the gasification industry—the updated worldwide gasification database 2014 update. In: Gasification Technologies Conference, Washington, DC

International Biochar Initiative (2022) Carbon gold: finding a market niche with a pilot scale biochar kiln. https://biochar-international.org/carbon-gold-finding-a-market-niche-with-a-pilot-scale-biochar-kiln/. Accessed 22 Mar 2024

International Energy Agency (IEA) (2020) Global biofuel production in 2019 and forecast to 2025. https://www.iea.org/data-and-statistics/charts/global-biofuel-production-in-2019-and-forecast-to-2025. Accessed 24 Oct 2023

International Energy Agency (IEA) (2021) Transport biofuels. https://www.iea.org/reports/renewables-2020/transport-biofuels. Accessed 4 Dec 2023

International Energy Agency (IEA) (2022) Global hydrogen review 2022, Paris. https://www.iea.org/reports/global-hydrogen-review-2022. Accessed 5 Nov 2023

International Energy Agency Bioenergy (IEA) (2023) Commercial status of direct thermochemical liquefaction technologies, IEA Bioenergy: Task 34. https://www.ieabioenergy.com/wp-content/uploads/2023/07/WP3.3-DTL-Final.pdf. Accessed 19 April 2023

International Energy Agency-Energy Technology System Analysis Programme (IEA-ETSAP) (2013) Biogas and bio-syngas production. Technology Brief P11. https://iea-etsap.org/E-TechDS/PDF/P11_BiogasProd_ML_Dec2013_GSOK.pdf#:~:text=Bio-SNG%20production%20costs%20are%20estimated%20in%20the%20range,10-15%25%20over%20the%20next%2010%20to%2015%20years. Accessed 19 Oct 2023

Khanna R, Li K, Wang Z, Sun M, Zhang J, Mukherjee PS (2019) Biochars in iron and steel industries. In: Jeguirim M, Limousy L (eds) Char and carbon materials derived from biomass: production, Characterization and Applications. Elsevier Inc., Mulhouse, pp 429–446

Kosakowski W, Bryszewska MA, Dziugan P (2020) Biochars from post-production biomass and waste from wood management: analysis of carbonization products. Mater (Basel) 13(21):4971

Lee J, Kim K-H, Kwon EE (2017) Biochar as a catalyst. Renew Sustain Energy Rev 77:70–79

Li H, Jiang L-B, Li C-Z, Liang J, Yuan X-Z, Xiao Z-H, Xiao Z-H, Wang H (2015) Co-pelletization of sewage sludge and biomass: the energy input and properties of pellets. Fuel Process Technol 132:55–61

Market Research (2022) Global plastic pyrolysis oil market research report 2022—market size, current insights and developments trends. https://www.marketresearch.com/Maia-Research-v4212/Global-Plastic-Pyrolysis-Oil-Research-32741395/. Accessed 5 June 2023

Market Research (2024) Fischer-Tropsch synthetic paraffinic kerosene—global market share and ranking, overall sales and demand forecast 2024–2030. https://www.marketresearch.com/QYResearch-Group-v3531/Fischer-Tropsch-Synthetic-Paraffinic-Kerosene-36537496/. Accessed 10 Apr 2024

Markets and Markets (2022) Biomass gasification market by source (agricultural, forest, animal, municipal), gasifier technology (fixed ben, fluidized bed, entrained flow), applications (power, chemicals, hydrogen, transportation, ethanol, biochar), region forecast to 2027 report. https://www.researchandmarkets.com/reports/5701289/global-biomass-gasification-market-by-source. Accessed 7 May 2023

Mensah KE, Damnyag L, Kwabena NS (2022) Analysis of charcoal production with recent developments in Sub-Sahara Africa: a review. Afr Geogr Rev 41(1):35–55

Michaels T, Krishnan K (2019) Energy recovery council. Directory of waste to energy facilities. http://energyrecoverycouncil.org/wp-content/uploads/2019/01/ERC-2018-directory.pdf. Accessed 20 Apr 2024

Mohammadi A (2021) Overview of the benefits and challenges associated with pelletizing biochar. Processes 9(9):1591

Mukherjee C, Denney J, Mbonimpa EG, Slagley J, Bhowmik R (2020) A review on municipal solid waste-to-energy trends in the USA. Renew Sustain Energy Rev 119:109512

Mulligan C, Strezov L, Strezov V (2015) Pyrolysis of biomass. In: Strezov V, Evans T (eds) Biomass processing technologies. CRC Press, pp 123–154

Nabukalu C, Reto G (2019) Charcoal as an energy resource: global trade, production and socioeconomic practices observed in Uganda. Resources 8(4):183

Nakamya M (2022) How sustainable are biofuels in a natural resource-dependent economy? Energy Sustain Dev 66:296–307

Njenga M, Kirimi M, Koech G, Wanjira EO, Muriuki J, Sola P, Bourne M, Siko I, Mendum R (2023) Improvements in charcoal production and the environmental implications: Potential for the invasive *Prosopis juliflora* in Kenya. Resour Conserv Recycl Adv 19:200181

NOVA Chemicals Corporation (2022). Advance commercialization of made-in-Alberta Chemical Recycling Technology. https://www.novachem.com/media-center/news-releases/nova-chemicals-and-enerkem-advance-commercialization-of-made-in-alberta-chemical-recycling-technology/. Accessed 3 Jul 2023

Ofori-Boateng C (2022) Algal biorefining: resource expenditure and exergo-environmental sustainability. Nova Science Publishers, New York, p 296

Oppong Mensah N, Addo S, Sumbayi SD, Osei-Gyabaah AP, Nakuja T, Anang SA (2024) Determinants of charcoal production and marketing in the Mankranso forest district in the Ashanti region of Ghana. Heliyon 10(1):e23800

Owens EM, Szczepkowski J (2010) Advancements in grate cooling technology. In: Proceedings of the 18th annual north American waste-to-energy conference, NAWTEC, Orlando, Florida, USA

Pennise D, Smith KR, Kithinji JP, Rezende ME, Raad TJ, Zhang J, Fan C (2001) Emissions of greenhouse gases and other airborne pollutants from charcoal-making in Kenya and Brazil. J Geophys Res Atmos 106:24143–24155

Qian K, Kumar A, Zhang H, Bellmer D, Huhnke R (2015) Recent advances in utilization of biochar. Renew Sustain Energy Rev 42:1055–1064

Renewable Energy World (REW) (2007). https://www.renewableenergyworld.com/baseload/bioenergy/range-fuels-signs-agreement-with-u-s-doe-breaks-ground-on-commercial-cellulosic-ethanol-plant-50495/#gref. Accessed 16 Dec 2023

Research Nester (2024) Chemicals and advanced materials, Fischer-Tropsch synthetic paraffinic kerosene market. https://www.researchnester.com/reports/fischer-tropsch-synthetic-paraffinic-kerosene-market/5777. Accessed 30 Apr 2024

Robb S, Joseph S, Abdul Aziz A, Dargusch P, Tisdell CA (2020) Biochar's cost constraints are overcome in small-scale farming on tropical soils in lower-income countries. Land Degrad Dev 31:1713–1726

Sasol (2023). https://www.sasol.com/sasol-and-topsoe-accelerate-growth-sustainable-aviation-fuel-saf. Accessed 11 Mar 2024

Sikarwar VS, Ming Z, Fennell PS, Shah N, Anthony EJ (2017) Progress in biofuel production from gasification. Prog Energy Combust Sci 61:189–248

Smith S (2021) Drax acquires Pinnacle Renewable Energy Inc. Energy Global Magazine. https://www.energyglobal.com/bioenergy/15042021/drax-acquires-pinnacle-renewable-energy-inc/. Accessed 10 Nov 2023

Spherical Insights LLP (2022) Global Fischer-Tropsch (FT) market size to grow. https://finance.yahoo.com/news/global-fischer-tropsch-ft-wax-090000995.html?fr=sycsrp_catchall. Accessed 19 Feb 2024

Srivastava N, Singh R, Singh P, Ahmad I, Singh RP, Rai AK, Asiri M, Gupta VK (2023) Recent advances on lignocellulosic bioresources and their valorization in biofuels production: challenges and viability assessment. Environ Technol Innov 29:103037

Statista (2022a) Global energy consumption by energy source 1990–2050. https://www.statista.com/statistics/222066/projected-global-energy-consumption-by-source/. Accessed 27 Sep 2023

Statista (2022b) Global $CO_2$ emissions 2018–2050. https://www.statista.com/statistics/263980/forecast-of-global-carbon-dioxide-emissions/. Accessed 27 Sep 2023

Statista (2024) Production of wood charcoal worldwide from 2000 to 2022. https://www.statista.com/statistics/1455987/wood-charcoal-production-globally/. Accessed 8 Mar 2024

Themelis N, Mussche C (2013) Municipal solid waste management and waste-to-energy in the United States, China and Japan. In: 2nd international academic symposium on enhanced landfill mining, Houthalen-Helchteren, pp 1–19

Thengane SK, Kung K, Hunt J, Gilani HR, Sokhansanj S, Jim Lim C, Sanchez DL (2021) Market prospects for biochar production and application in California. Biofuels Bioprod Biorefin. https://bof.fire.ca.gov/media/oobbtosm/thengane_2021_ca_biochar_market-002-_ada.pdf

Trend Economy (2024) World merchandise exports and imports by commodity. https://trendeconomy.com/data/commodity_h2/4402. Accessed 19 Mar 2024

United States Department of Energy (US DOE) (2019) DOE Bioenergy Technologies Office (BETO) 2019 project peer review, red rock biofuels project update. https://www.energy.gov/sites/prod/files/2019/04/f61/Woody%20Biomass%20Biorefinery%20Capability%20Development_EE000DPA2.pdf. Accessed 11 Dec 2023

Virtue Market Research (2023) Charcoal market research report—forecast (2024–2030). https://virtuemarketresearch.com/report/charcoal-market. Accessed 17 Feb 2024

Waste Management World Magazine (2013) https://waste-management-world.com/artikel/jbi-finishes-3rd-waste-plastics-to-oil-processor-in-new-york/. Accessed 20 Dec 2023

World Bioenergy Association (2020) Global bioenergy statistics. https://www.worldbioenergy.org/uploads/201210%20WBA%20GBS%202020.pdf. Accessed 15 Jun 2023

Ye L, Peng Z, Wang L, Anzulevich A, Bychkov I, Kalganov D et al (2019) Use of biochar for sustainable ferrous metallurgy. JOM 71(11):3931–3940

Zhu G, Yang L, Gao Y, Xu J, Chen H, Zhu Y, Wang Y, Liao C, Lu C, Zhu C (2019) Characterization and pelletization of cotton stalk hydrochar from HTC and combustion kinetics of hydrochar pellets by TGA. Fuel 244:479–491

# Part II
# Thermochemical Biomass Conversion Technologies

# Chapter 3
# Current Thermochemical Biomass/Waste Conversion Pathways

**Abstract** This chapter discusses the current advanced technologies for thermochemical conversion of waste materials into biofuels, bioenergy, and biochemicals by providing in-depth awareness of the position these technologies play in the global shift toward circular economy and sustainability. Thermochemical conversion processes are found to be the most sustainable ways of transforming all kinds of waste like solid (e.g., food waste, lignocellulosic waste, construction waste, plastic waste) and liquid waste (e.g., sewage sludge) into bio-based products compared to biochemical, biological, and chemical conversion technologies. Since waste may contain diverse components especially when they are not sorted at source but damped as a mixture at landfills, thermochemical waste conversion technologies become innovative and sustainable solutions for managing waste whilst simultaneously generating bioproducts. Thermochemical conversion technologies like pyrolysis and gasification are discussed, and recent advancements are explored in this chapter.

**Keywords** Thermochemical conversion · Waste · Pyrolysis · Gasification · Combustion · Hydrothermal treatment · Waste-to-energy (WtE) · Plasma treatment

## 3.1 Introduction

With the world's current waste generation rate exponentially increasing whilst recycling and reuse rates are low, new technologies are required for waste transformation into eco-friendly and cost-effective products. Biofuels and biochemicals produced from waste materials like plastics, food waste, and agricultural residues have gained tremendous recognition as sustainable bioproducts to replace petroleum and fossil-based products. About 90% of waste materials generated in the world are collected as mixtures of waste at source and it becomes costly to sort them into various categories for efficient recycling. Based on the physicochemical characteristics of the feedstock used to produce eco-friendly bioproducts, suitable technologies can be adopted for sustainable production. However, most of the waste conversion technologies cannot efficiently transform a mixture of wastes (e.g., a mixture of plastics and food waste)

 
C. Ofori-Boateng, *Sustainability of Thermochemical Waste Conversion Technologies*,
https://doi.org/10.1007/978-3-031-64342-2_3

into value-added bioproducts. Nonetheless, thermochemical conversion technologies offer sustainable alternatives to waste recycling especially when waste sorting is costly and there is no option but to use mixed waste.

Thermochemical conversion processes such as pyrolysis, torrefaction, combustion, gasification, and hydrothermal treatments are popular and efficient methods used to valorize waste materials into value-added bioproducts. Unlike other biomass conversion processes, thermochemical conversion technologies transform the entire feedstock into products in a short time using high temperatures to aid the decomposition process. Figure 3.1 shows the basic process conditions necessary for the conversion of wastes into bioproducts via the common thermochemical conversion technologies. In order to achieve sustainable production of value-added bioproducts from waste materials via thermochemical means, it is imperative to continuously improve the process. Recent advancements in renewable energy introduce improved thermochemical conversion technologies which are discussed in this chapter. Resource expenditure in some thermochemical conversion technologies is described in detail elsewhere (Ofori-Boateng 2022).

## 3.2 Combustion

Combustion is an exothermic chemical reaction whereby waste material (called the fuel) is burnt at 900–1200 °C in the presence of oxygen (called the oxidant) to decompose the organic and inorganic matter into stable valuable products (called oxidized products) like biochar, syngas, etc. and residues like ash (Jegla et al. 2010; Williams 2013). The hydrocarbons in the feedstock react with oxygen in the air to form carbon dioxide, heat, and water according to Eq. 3.1:

$$\begin{array}{ccccccccc} \text{Hydrocarbon} & + & \text{oxygen} & \rightarrow & \text{Water} & + & \text{Carbon dioxide} & + & \text{heat} \\ C_xH_y & + & O_2 & \rightarrow & H_2O & + & CO_2 & + & \text{Heat} \end{array} \tag{3.1}$$

Figure 3.2 shows the flow diagram of electricity generation from the combustion of waste. The generation of energy from combustion of waste begins with waste collection and preparation. The waste is sorted by an overhead crane which is then lifted into the combustion chamber to be burnt.

The heat generated during the process is used to heat water into steam, which is then sent to a steam turbine and generator for electricity production. The fly ash is then collected and filtered at a landfill site which can remove all particulate matter in the gas stream. The captured fly ash particles are then sent to ash discharger via enclosed conveyer systems where they are mixed with water to reduce dust and then mixed with the bottom ash products.

At the end of the combustion reaction, combustible particulate matter called bottom ash (about 15–25 wt% of input waste material) is formed in the combustion chamber (EPA 2024). Also, some of the combustible particles are emitted into

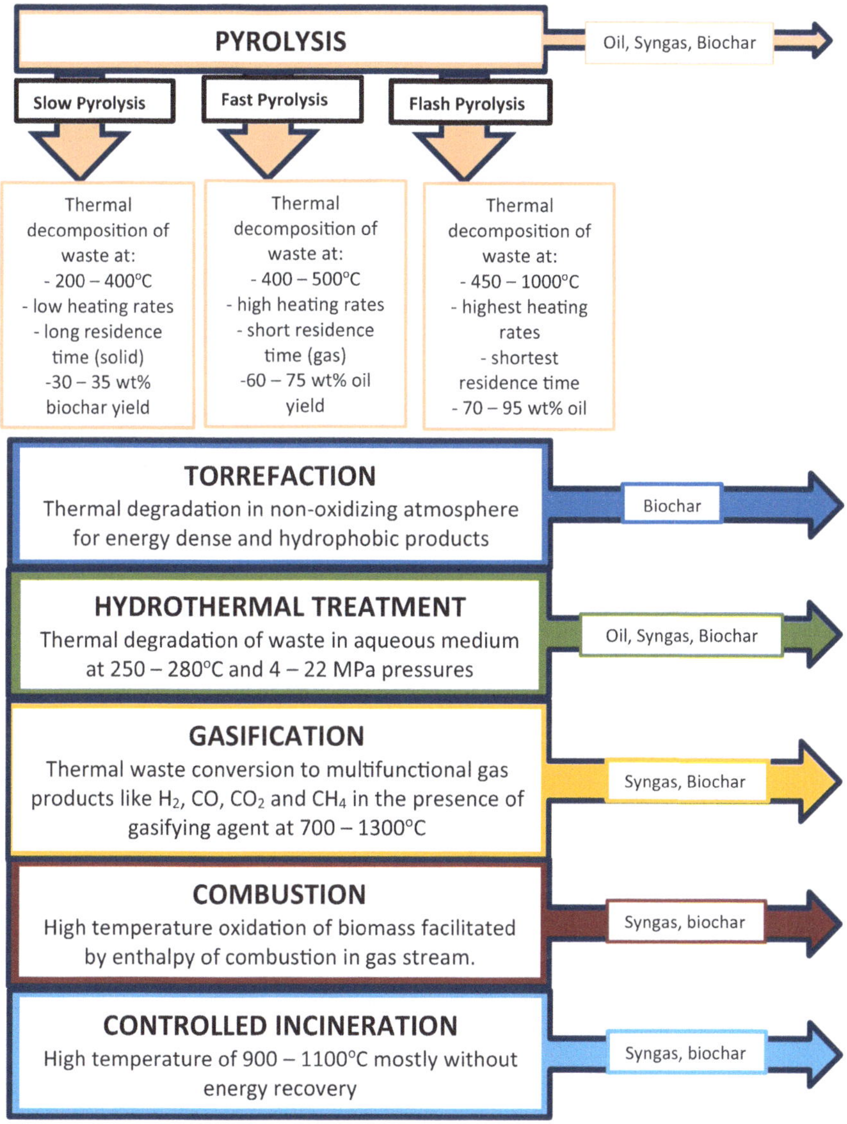


**Fig. 3.1** Summary of common thermochemical waste conversion technologies

the atmosphere together with exhaust gas, and this pollutant is called fly ash. Fly ash forms about 10–20 wt% of the total ash generated by the combustor (EPA 2024). Other products of combustion include oxides of sulfur ($SO_x$) and nitrogen oxides ($NO_x$).

The common technologies for combustion of waste are mass burn facilities, refuse derived fuel systems, and modular systems. Mass burn facilities process waste (either

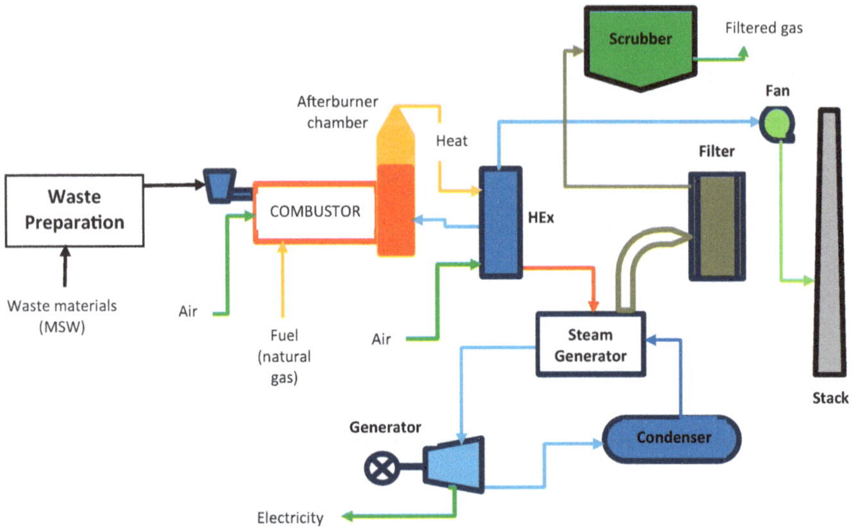


**Fig. 3.2** Process flow diagram of waste combustion to bioenergy

sorted, prepared, or not) in a single combustion chamber under excess air which promotes uniform mixing to facilitate turbulence for efficient combustion. Modular systems are mobile, much smaller, and compact than mass burn units, and they burn unprocessed MSW (municipal solid waste). Refuse derived fuel systems on the other hand use mechanical means to shred the waste, separate the components, and pretreat it before being sent into the combustor. Currently, according to the Environmental Protection Agency (EPA), the United States operates close to 76 combustion facilities that recover energy from MSW which generates about 550 kWh of electricity per ton of waste (EPA 2024).

## 3.3 Incineration

Incineration[1] is an exothermic process which involves the burning or combustion of solid or liquid waste or biomass in the presence of oxygen at high temperatures of 900–1200 °C and high pressure (National Research Council 2000). The generated hot flue gas stream can be fed into a steam turbine unit to produce electricity and heat. Incineration systems can recover up to 545 kWh of energy from a ton of waste incinerated (Zaman 2010). Unlike combustion, during incineration, combustible compounds

[1] Incineration is different from flaring. Flaring is a combustion control process where the gases and vapors are channeled through pipes to a specific place and burnt in an open flame into the atmosphere via series of specially designed tools to destroy the gases.

in the gas generated are usually released into the atmosphere without capturing it for energy production.

Controlled incineration has become one of the main thermochemical conversion technologies to transform waste materials especially Municipal Solid Waste (MSW) into bioenergy (heat and electricity) due to its high degradation efficiency and contaminant removal potentials (Palacio et al. 2018). Incineration reduces the amount or volume of solid waste (by about 70–90%) and its environmental impacts, thus most industrialized countries where MSW generation is high have adopted incineration as a better alternative to landfilling. Like gasification and combustion, incineration can be used to recover the energy value of waste materials whilst pyrolysis and hydrothermal treatment recover the chemical value of waste materials. Other advantages of incineration include:

- High flexibility of the type of waste it can handle. Basically, incineration systems can handle all kinds of waste in large quantities.
- Vast applications of by-products. Bottom ash can be used as an eco-friendly building material.

Controlled biomass or waste incineration systems which operate in cogeneration mode (for biochar and biogas) and combined heat and power systems have low efficiency (20–25% and 25–35%, respectively) compared to coal incineration at similar conditions (about 40% efficiency) (Di Maria et al. 2016). In both scenarios, as much as incineration reduces the amount of solid waste hence minimizing land pollution, more than 50% of the heat generated is wasted into the environment rendering the technology prone to air pollution and energy inefficient. Hazardous air pollutants (e.g., furan, mercury, cadmium, etc.) released into the environment during incineration affect the performance of steam cycle negatively. Controlled waste incinerators therefore have pollution control devices like afterburners, scrubbers, and filters to remove particulate matter, acid gases, carbon monoxide, etc., to remove as many pollutants as they are released into the environment.

Also, the recovery of energy via new and advanced technologies like pyrolysis or integrated systems for energy capture can render incineration more sustainable. Also, Liao et al. (2023) reported that about 25% oxygen-enriched incineration could improve combustion efficiency and minimize air pollutants especially when using high moisture content MSW (>25 wt% moisture). Thus, high oxygen concentrations would inhibit the impact of water (or moisture) in the feedstock on pollutant emissions hence improving MSW Incineration technologies (Liao et al. 2023). In a similar report, excess oxygen injection into the incinerated feedstock helped improve the ignition and complete combustion process through the enhancement of the flame temperature in the boiler and reduction of the excess air coefficient (Ma et al. 2019). This strategy can increase fuel availability and drastically reduce emissions into the atmosphere. Moško et al. (2018) also concluded in an incineration system investigation on emissions that increasing oxygen concentration reduces the $SO_x$ and $NO_x$ emissions because of the presence of water vapor.

Figure 3.3 shows the flow diagram of a typical waste incineration. The first stage in controlled waste incineration is feed preparation. Feed preparation involves the

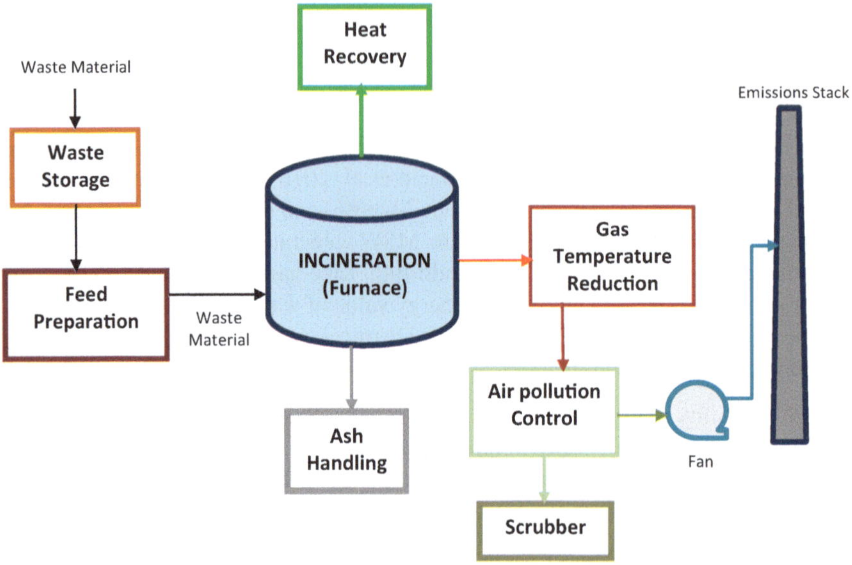


**Fig. 3.3** Process flow diagram of waste incineration

removal of over-size non-combustible materials, shredding, blending of compatible waste materials, etc. Crane, bucket, hopper, pneumatic conveyors, and ram feeders are devices which aid the charging of the feed into the combustor. Since waste preparation and feeding have a potential for fire hazards and emissions, it must be properly managed to ensure good furnace combustion performance.

In the furnace, the waste material is dried and devolatilized after which the volatiles and soot are combusted followed by the final combustion of the solid carbonaceous residues. When the waste enters the incinerator, preheated air (or heat radiation from incinerator walls) reduces the moisture content of the waste which then facilitates the thermal degradation of the organic matter at 200–750 °C (Williams 2013). As a result, combustible gases, and volatile organic matter (e.g., methane, hydrogen, high molecular weight hydrocarbons, etc.) are generated at 425–550 °C depending on the composition of waste (Williams 2013). The volatile matter produced is then burnt at 850–1200 °C for a short residence time in the combustion chamber situated above the grate to release gases (National Research Council 2000; Williams 2013). In order to prevent the formation of slag by fusion of ash residues, the combustion temperature must not exceed 1200 °C. Again, to avoid incomplete burning, the generated vapors and gases should not be held in the combustion chamber for less than 2–4 s whilst the supply of air through the nozzles above the grate is maintained (Williams 2013).

Combustion gases from the furnace (are released in a form of flue gas at 1000–1200 °C), are cooled in gas temperature reduction units such as waste-heat boilers and direct-contact water spray quenches (National Research Council 2000). The main solid residues generated by incineration are fly ash, scrubber water, bottom ash, and other waste streams. Bottom ash is the remains of the waste which did not burn on the grate of the furnace during combustion. Bottom ash composition depends on the type of waste used. MSW bottom ash comprises large pieces of glass, ceramic, char, and inorganic fine particles. After incineration, bottom ash can be removed into a quench pit below the burnout section of the grate.[2] Fly ash, on the other hand, is the solid condensable vapor suspending in combustion gases which leaves the combustion chamber. Fly ash consists of organic chemicals, volatile metals, and condensed acids (National Research Council 2000). Scrubber water is a slurry that is obtained after wet scrubbing of the generated gases from the furnace, and it consists of salts, excess lime, and contaminants like condensed organic vapors. In controlled incineration like combustion where heat is captured for energy generation, blow-down water from the heat recovery boilers, and condensed metals from parts of the flue gas system are considered waste streams and must therefore be appropriately disposed of (National Research Council 2000).

The incinerator has a flame zone where the combustion occurs, and this chamber must be hot enough to decompose all the organic and inorganic matter in the waste hence facilitating the reactions between volatile compounds of the waste the oxygen and nitrogen in the air. Carbon and oxygen react producing carbon dioxide whilst hydrogen and oxygen react producing water vapor during incineration. The evolution of carbon monoxide and other carbon containing particulate matter is caused by incomplete burning in the furnace. In the incinerator, hydrogen also reacts with organically bound chlorine to form hydrogen chloride (HCl). Many series of other reactions also occur in the furnace during combustion releasing some compounds such as nitrogen oxides, metal oxides, metal vapors, etc. Air pollution (from particulate matter and acid gas emissions) in incineration facilities are controlled and minimized via techniques and devices such as fabric filters, electrostatic collectors (and inertial-impaction collectors) and ionizing wet scrubbers (and venturi scrubbers), precipitators (National Research Council 2000), without which the gases may be released into the atmosphere directly which is environmentally unsustainable.

Incomplete burning in the furnace (mostly near the walls) may be due to improper design, operation, turbulence of the gas mixture in the furnace, and gas residence time at the specified temperature. In order to achieve effective combustion in the furnace, the gas stream must attain the appropriate temperature for the specified time whilst ensuring uniform mixing of the fuel and oxygen. The temperature in the furnace is mostly maintained at a steady point by controlling the quantity of materials fed into the furnace whilst varying the amount of excess air so that the rate of heat evolution would stay within the targeted range. Thus incineration plants must incorporate sustainable ways of reducing emissions during the combustion process. For instance, there are new design features of incinerators which help maintain appropriate temperature, residence time, and turbulence in the furnace to improve combustion efficiency. Such features include pretreatment methods such as oxygen enrichment as well as high efficiency burners (National Research Council 2000).

[2] Grates are components of incineration system which transport the waste into the combustion zone. Advanced designs of grates in incinerators can grind or shred the waste into smaller pieces to increase its surface area for efficient burning.

**Table 3.1** Advantages and disadvantages of waste incineration (Williams 2013)

| |
|---|
| *Advantages* |
| • High waste reduction up to about 95 wt%<br>• Ash can be safely disposed of in landfills<br>• Sterile residue due to high-temperature incineration compact and can be done near the location of collection<br>• Does not generate methane gas unlike landfilling<br>• Possibility of energy recovery when controlled<br>• Best thermal treatment for hazardous, flammable, infectious, and toxic waste |
| *Disadvantage* |
| • Harmful emissions risks<br>• High capital investment and maintenance cost<br>• Inorganic residues generated may need further treatment before disposal<br>• Requires skilled labor for operation |

There are many different types of incineration furnaces including batch-fed, stationary refractory hearth, fixed-hearth incinerators (for medical wastes), rotary kiln, continuous feed, reciprocating grate designs, light-weight aggregate kilns, industrial boilers, halogen-acid recovery furnaces, sulfuric-acid regeneration furnaces, and waste-to-energy facilities with radiant waterwall furnaces and convective boilers (National Research Council 2000). The commonly used incinerator designs are liquid-injection furnaces and rotary kilns which are suitable for medical wastes and other hazardous waste materials (National Research Council 2000). Table 3.1 summarizes the advantages and disadvantages of waste incineration. The recent trend in incineration focuses on the upgrade of the waste material by prior pretreatment using energy densification methods like pelleting, briquetting, roller pressing, cuber, extruder, pressure agglomerators, etc.

## 3.4 Pyrolysis

Pyrolysis[3] is the decomposition of organic matter in feedstock or waste materials into different valuable components (such as bio-crude oil,[4] biochar,[5] and syngas[6]) using high temperature (between 300 and 900 °C) in the absence of oxygen (Palanivelrajan and Feroskhan 2023). The conditions of pyrolysis as well as the type of waste material (see Chap. 1) would affect products' yields and quality. The types of pyrolysis processes are extensively discussed elsewhere (Ofori-Boateng 2022).

[3] Pyrolysis is an endothermic process meaning it requires energy to break chemical bonds.

[4] Pyrolysis oil consists of furanic compounds, hydrocarbons, carboxylic acids, aldehydes, ketones, water, phenolic compounds, and traces of water depending on the type of feedstock used.

[5] Pyrolysis biochar consists basically of carbon and ashes.

[6] Pyrolysis syngas (non-condensable gas) consists of $CO_2$, CO, and $CH_4$.

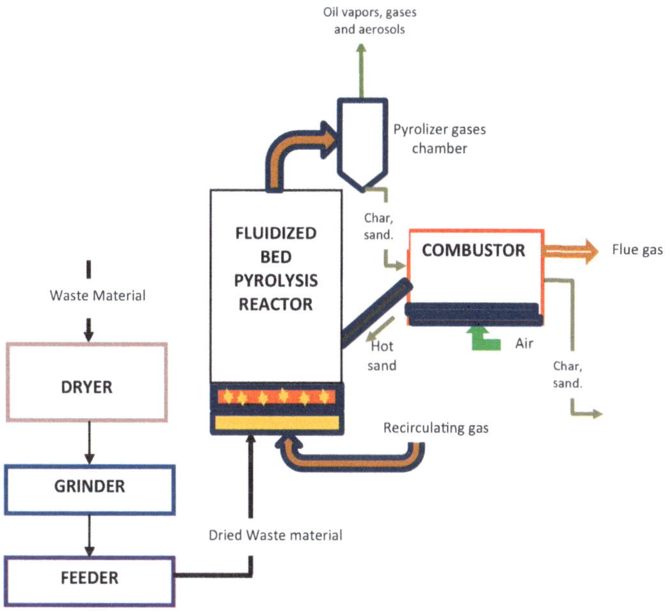


**Fig. 3.4** Process flow diagram of waste pyrolysis

Pyrolysis reaction occurs in two stages namely primary pyrolysis stage and secondary pyrolysis stage. During the first stage of pyrolysis, the feedstock is heated at low temperatures (<250 °C) which keeps increasing over time to 500 °C and over, where the products generated at this stage move into the secondary reaction stage as intermediate. During the secondary reaction, cracking,[7] reforming, dehydration, condensation, polymerization, charring, oxidation, and gasification occur where the chemistry is described in detail by Wang et al. (2017).

Pyrolysis oil, bio-syngas, and biochar can further be processed into biofuels, biochemicals, and biofertilizers for vast applications in the thermal plants and chemical and transportation industries. Figure 3.4 shows the process flow diagram of pyrolysis of waste into value-added bioproducts.

Pyrolysis temperature, presence or absence of catalysts, reactor type, heating rate, pressure, residence time, particle size of waste material as well as the reaction environment (medium or atmosphere) affect the yield and quality of pyrolysis oil produced via fast pyrolysis.

***Pyrolysis temperature*** is one of the most important factors that is extensively studied in pyrolysis because it affects the decomposition, dehydrogenation, isomerization, and polymerization reactions which intend to affect the oil yield and quality

[7] Cracking is the breakdown of alkanes into smaller molecules at high temperatures between 650 and 970 °C.

(Ibrahim 2020). High reaction temperatures facilitate the excitation of molecules to incite polymeric macromolecule changes which break long-chain polymers into short ones. This leads to the evolution of free radicals which aid the degradation of the polymers by depolymerization and side-group elimination reaction (Zhang et al. 2022). Increase in reaction temperature during pyrolysis leads to increase in oil yield and vice versa as evidenced in reported works (Chang 2023).

***Pyrolysis residence time and heating rate*** are critical factors that affect the yield of oil. Residence time is the time the feedstock or waste material stays in the reactor under pyrolysis conditions until it reaches the desired temperature. Residence time includes the time that the feedstock begins to heat up and the time the reaction is completed (Marathe et al. 2020). The heating rate on the other hand is the rate at which the reaction temperature in the pyrolysis reactor increases with formation of products. Since pyrolysis is a thermal decomposition process, heating rate and residence time affect the heat and mass transfer distributions within the reactor hence impacting the quality and quantity of pyrolysis oil. Generally, low-to-moderate pyrolysis heating rate (5–200 °C/min) and moderate-to-high residence time (15–240 min) with temperatures of 250–800 °C speed up the formation of high-quality and high pyrolysis oil yields (Singh et al. 2019; Riesco-Avila et al. 2020). On the other hand, formation of char becomes dominant in pyrolysis when the residence time is short and heating rate is high (Cahyono and Styana 2017).

***Pyrolysis operating pressure*** can be atmospheric (1 bar), in vacuum (<1 bar), and above atmospheric pressure (2–9 bar). Most pyrolysis reactions occur under atmospheric pressure due to the high cost of operation using vacuum and high pressures (Chang 2023). During pyrolysis, random-chain scission which yields pyrolysis oil does not proceed at increased pressure but end-chain β-scission which gives rise to vapor production in the reactor is pressure dependent and increases with increase in pressure (Murata et al. 2004). Also, for the formation of high yield biochar during pyrolysis, an increase in operating pressure is required for the formation of products resulting from dehydration, polymerization, and condensation reactions.

***Catalysts*** are compounds used in pyrolysis to facilitate the maximum decomposition at relatively low temperature and short residence time. Pyrolysis can proceed as catalytic (i.e., with a catalyst) or non-catalytic/thermal (i.e., without catalysts). Catalytic pyrolysis reduces the amount of heat utilized for the degradation of waste material. Catalytic pyrolysis accelerates isomerization and hence generates a high quantity of vapors (as fuel) compared to oil though the oil possesses better qualities like diesel ($C_{13}$–$C_{20}$), wax ($\geq C_{20}$), monomers (e.g., propylene, styrene, ethylene, etc.) and gasoline ($C_6$–$C_{12}$) (Lerici et al. 2015). According to Al-Salem et al. (2017), non-catalytic pyrolysis produces high yield bio-oil (and non-fuel type vapors) with high hydrocarbon content compared to catalytic pyrolysis. The choice of catalyst influences other pyrolysis factors such as temperature, reaction time, etc., for efficient process. Some types of catalysts used for pyrolysis of waste materials include silica-alumina catalysts, red mud, zeolite-based catalysts (e.g., HZMSM-5, home-made beta zeolite [H-BEA], etc.), fluid cracking catalysts (FCC), reforming catalysts, etc. Zeolite-based catalysts are found to improve pyrolysis by reducing the formation of impurities like ash, nitrogen, sulfur, etc., in the oil, as well as increase the rate of cracking generating more gases than liquids (Syamsiro et al. 2014).

Pyrolysis reactor type and configuration are found to affect the generation of pyrolysis products. Some commonly used pyrolysis reactors include tubular, continuous flow, fixed bed, fluidized bed, continuous stirred tank, auger, and conical spouted bed reactors (Park et al. 2022; Chang 2023). Pyrolysis reactors can be operated in simple batch or semi-batch mode designs which have the capability to produce a vast variety of products using different reactants and operating conditions (Sekar et al. 2022). On the other hand, continuous flow reactors such as fluidized bed reactors have lower cost of labor and are efficient for commercial scale catalytic and non-catalytic/thermal pyrolysis of waste. Nonetheless, fixed bed, continuous stirred tank, and auger reactors are easy to operate due to the limited amount of carrier gas needed as well as no bed materials which may render the operation simple and compact (Al-Salem et al. 2017). Conical spouted bed reactors are costly but efficient for adhesive feedstocks like plastic wastes compared to fluidized bed reactors (Sekar et al. 2022). Table 3.2 summarizes the pros and cons of the commonly used reactors for waste pyrolysis.

### 3.4.1 Pyrolysis of Plastic Waste

Plastic waste generation is one of the challenging issues confronting the world as it is found everywhere on earth including the oceans and can cause health concerns to humans as well as destroy wildlife and the ecosystem at large. In as much as plastic products can take close to 10 years to expire, it takes about 480–500 years to decompose depending on the characteristics and disposal method (Alves 2024; OECD 2022). Currently, the world generates about 450 million tons of plastic waste annually of which only about 10–20% is recycled (OECD 2022). In Europe for instance, for a total of about 58.5 million tons of plastic waste generated annually, only about 9.4% and 0.9% are mechanically[8] and chemically recycled, respectively (Qureshi et al. 2020), and the rest is incinerated and damped at landfills. The United States and United Kingdom are among the top generators of plastic wastes in the world collectively contributing about 204 kg/capita per year (Law et al. 2020). Although China is the largest producer and consumer of plastics in the world, its plastic waste generation per capita is among the lowest at about 15.7 kg/year because of the implementation and enforcement of the National Sword Policy which bans the import of plastic wastes and single use of plastic straws (Wen et al. 2021; Law et al. 2020; Zhang 2017; Zhang 2021).

Thermochemical recycling of plastic waste is the best way to convert waste into value-added bioproducts which would play a pivotal part in the transition towards

[8] Mechanical waste recycling comprises waste sorting, cleaning, melting, and forming into granules and other new products.

**Table 3.2** Advantages and disadvantages of types of pyrolysis reactors (Sekar et al. 2022)

| Type of pyrolysis reactor | Advantage | Disadvantage |
|---|---|---|
| Batch reactors | • High feedstock conversion<br>• Easy to operate | • Difficulty in scaling up<br>• Time consuming to clean and maintain<br>• Products vary from batch to batch<br>• High cost of labor |
| Semi-batch reactors | • Flexibility of adding and removing products over time<br>• Easy to operate | • High cost of labor<br>• Difficulty in scalability<br>• Inefficient with catalytic pyrolysis<br>• Not suitable for commercial scale production |
| Fluidized-bed reactors | • Uniform heat distribution<br>• High degree of cracking<br>• Easy accessibility and recovery of catalysts without unloading<br>• Better mixing of catalysts with fluid<br>• Flexible and scalable<br>• Better consistency in product mix | • Char-bed material separation problem<br>• High cost of operation |
| Fixed bed reactors | • Simple and compact design | • Low heating rate<br>• Inconsistent temperature distribution in reactor<br>• Catalyst access difficulty<br>• Not suitable for non-uniformly mixed feedstock |
| Continuous Stirred Tank Reactor (CSTR) | • Better temperature control<br>• Easy to operate and design<br>• Uniform heat distribution | • Low conversion rate per volume<br>• Non-uniform mixing<br>• High maintenance cost<br>• High labor cost |
| Auger reactors | • Can be used for both fast and slow pyrolysis<br>• Less solid fraction attrition<br>• Pyrolysis gas is less diluted<br>• Versatile during installations<br>• Suitable for mobile applications<br>• Used for all kinds of waste materials<br>• Scalable to commercial scale | • Mechanical wear and tear of parts due to high temperature<br>• Heat transfer difficulties on a large scale<br>• Poor mixing at radial direction<br>• High residence time |

(continued)

**Table 3.2** (continued)

| Type of pyrolysis reactor | Advantage | Disadvantage |
|---|---|---|
| Conical spouted bed reactors (CSBR) | • Uniform mixing<br>• Suitable for feedstock variety mixture<br>• Better heat transfer between phases | • Difficult catalyst feeding<br>• Difficult product and entrainment removal<br>• Complicated design<br>• High operating cost |
| Rotary kiln reactors | • Better mixing of feedstock<br>• Best for variety of feedstock mixture<br>• Easy maintenance | • The process is slow<br>• High formation of char<br>• Low heating rate |
| Tubular reactors | • Simple and safe to use<br>• Low labor cost<br>• Stable operation<br>• Short residence time<br>• Suitable for both catalytic and non-catalytic pyrolysis | • Susceptible to equipment corrosion<br>• The control system is complex<br>• Feedstock must be pretreated<br>• Small area of tubes makes it suitable for small-scale production at a time |
| Microwave pyrolysis reactor | • Efficient heat transfer<br>• Suitable for large amount of feedstock<br>• Uniform heat distribution<br>• High heating rate<br>• Simple and compact | • High operating cost |

circular economy. Hazardous chemicals like brominated flame retardants, phthalates, and some hydrocarbons in plastic wastes are effectively removed during thermochemical treatment processes. Plastic waste is a valuable feedstock for fuel and energy production via pyrolysis due to its reportedly high calorific value of about 41–47 MJ/kg compared to coal (about 25–35 MJ/kg) and wood/forest residues (15–19 MJ/kg) (Dincer et al. 2018; Günther et al. 2012). However, the combustion of plastic pyrolysis oil as fuel may emit exhaust gases like CO and $CO_2$ based on the hydrogen-to-carbon ratio (high H/C ratio in the oil may lead to less emissions) (Gopalakrishnan et al. 2019). The pyrolysis of plastic waste has been reported to produce a variety of compounds with carbons ranging from six to forty (i.e., $C_6$ to $C_{40}$) (Chang 2023). Generally, depending on the process conditions and plastic waste composition, plastic fast pyrolysis may generate variety of products with different chemical composition such as:

- Liquid plastic oil consisting of aliphatic compounds, mono- and polyaromatic compounds which can be used directly in boilers for energy generation or as a raw material to produce monomers, biodiesel, bio-gasoline, nanotubes, etc.

- Waxy plastic oil made up of long-chain alkanes and alkenes (above $C_{20}$[9]) with high boiling points. Waxy plastic oil, unlike liquid plastic oil, may require additional processing like catalytic cracking before they can be used in boilers and automobile engines as fuels.
- Non-condensable gases made up of methane, hydrogen, carbon monoxide, carbon dioxide which can be recycled back into the reactor as a source of heat.
- Carbon-rich solid biochar which can be used for briquettes and pellets production.

Plastic Pyrolysis oil is a highly dense and viscous dark brown liquid with higher high heating value (HHV) closer to petroleum fuels (Peng et al. 2022) and has many applications in the chemical and transportation industries. Plastic pyrolysis oil has high stability and lower acidity thus purification before applications in boilers and burners is not a major requirement compared to pyrolysis oil from other feedstocks (Hu and Gholizadeh 2020). By use of high-resolution gas chromatography (GC) like on-line comprehensive two-dimensional chromatography (GCxGC), the complete constituents of pyrolysis oil from plastic waste mixture may be determined (Dao et al. 2021; Zanella et al. 2023) for informed decision for their applications.

On a laboratory scale, the fast pyrolysis of plastic wastes at 250–500 °C yielded about 73–99 wt% plastic oil at a heating rate of 10–20 °C/min for 90–150 min in a nitrogen environment without a catalyst (Hariadi et al. 2021; Sonawane et al. 2017; Singh et al. 2019). Prurapark et al. (2020) reported a different trend in pyrolysis plastic oil where about 22.5 wt% yield was obtained at 400 °C. Nonetheless, at similar pyrolysis conditions with an HZSM-5 zeolite catalyst and high reaction temperature of 550 °C, aromatic enriched plastic oil yield was lower at about 36–44 wt% (Inayat et al. 2022). Lower plastic pyrolysis oil yields in the range of 20–69 wt% have been reported at elevated reaction temperatures of 500–750 °C with or without catalyst at residence times of 30–60 min (Akubo et al. 2019; Park et al. 2022). According to Zhou et al. (2016), high temperatures during plastic pyrolysis lead to the production of toxic polycyclic aromatic hydrocarbons (PAHs) through the Diel-Alder[10] reaction (polyolefins formation), hydrogen-abstraction-acetylene-addition mechanism (aromatic hydrocarbons polymers), and chain scission process which together pose health hazards when emitted into the environment. Nevertheless, the cracking of waxy plastic oil into less viscous liquid is accelerated by high pyrolysis temperatures (Cahyono and Styana 2017). For optimum plastic pyrolysis oil production, the optimum temperature must range from 380 to 730 °C and 350 to 500 °C with and without catalysts, respectively, for mixed plastic waste (Chang 2023).

At 500 °C, Singh et al. (2019) pyrolyzed mixed plastic waste at a very high heating rate of 1,200,000 °C/min and obtained plastic oil yield of about 7 wt% compared to 82 wt% oil yield at 20 °C/min using the same pyrolysis condition conditions. Also,

---

[9] $C_{20}$ or carbon-20 or $^{20}C$ is an isotope of carbon where the carbon atoms in chain is 20. Alkanes with C20 is called Eicosane ($C_{20}H_{42}$).

[10] Diel-Alder reaction is an organic reaction where a conjugated diene (molecule with two alternating double bonds) and a dienophile (an alkene) are converted into olefins in a one-step reaction in the presence of heat.

when the residence time for mixed plastic waste pyrolysis (at 500 °C) was increased from 10 to 30 min, the plastic oil yield also increased from about 49 wt% to about 65 wt% (López et al. 2011). The effect of medium residence time and low-to-moderate heating rate on plastic waste pyrolysis at low-to-moderate temperature is attributed to the steady degradation of the waste which makes the generated vapors remain in the heating zone for a long time as opposed to rapid decomposition at high heating rates, high temperatures and short residence time where volatiles dissociate by chain β-scission and molecular hydrogen transfer yielding more monomers and short-chain hydrocarbons hence low plastic oil quality and yield (Chang 2023; Soni et al. 2021).

In a continuous stirred tank reactor (CSTR), the increase in plastic pyrolysis pressure from 1 to 8 bar resulted in increased gas yield from 6 to 13 wt% at 410 °C compared to decreased gas yield of 4 to 6 wt% at 440 °C (Murata et al. 2004). This shows that pressure is temperature dependent as evidenced in other reports where increased pressure from 6 to 51 bar decreased oil yield from 89 to 95 wt% as the gas production increased at 380 °C (Cheng et al. 2020).

Catalytic pyrolysis of plastic waste mix at 460 °C using 10% zeolite catalyst (ZSM-5) at a heating rate of 20 °C/min for 0–15 min, bio-oil, syngas, and biochar yields were 56.9 wt%, 40.4 wt%, and 3.2 wt%, respectively (López et al. 2011). The catalytic pyrolysis oil yield (56.9 wt%) was comparatively lower than non-catalytic/ thermal pyrolysis oil (79.3 wt%). The conclusion of Al-Salem et al. (2017) confirms the results of López et al. (2011) and other reports (Syamsiro et al. 2014; Aisien et al. 2021; Singh et al. 2020) that catalytic pyrolysis of plastic waste generates high yields of gases (40.4 wt%) compared to non-catalytic pyrolysis (17.7 wt% gas yield). Generally, catalytic pyrolysis of plastic wastes reduces the reaction temperature, residence time, and energy requirement as well as changes the pyrolysis pathway hence product selectivity to generate oil with high olefins, aromatics, and paraffinic contents (Peng et al. 2022). In some reports, however, high yields of plastic oil have resulted from catalytic pyrolysis of plastic wastes which indicates reduced secondary over-cracking of pyrolytic intermediate (Hakeem et al. 2018).

Cai et al. (2023) investigated the efficiency of catalytic pyrolysis of mixed plastic wastes using different kinds of clay catalysts (nanoclay, kaolin, hydrotalcite, and montmorillonite). Their results showed that montmorillonite catalyst facilitated the production of pyrolysis products at high yields (71 wt%) whilst kaolin and nanoclay promoted the evolution of gasoline and diesel range hydrocarbons in the oil, respectively, due to the difference in clay catalyst acidity contents.

Plastic pyrolysis oil is reported to have characteristics closer to those of fossil liquid fuels. For instance, plastic oil has lower API gravity (24–33) than fossil liquid fuels (38–55) which makes plastic oil denser (Chang 2023). Also, plastic oil viscosity is found to be in the range of diesel viscosity thus has similar properties. Table 3.3 compares the characteristics of petroleum crude oil, diesel, and gasoline to plastic pyrolysis oil.

**Table 3.3** Advantages and disadvantages of types of pyrolysis reactors (Ahmad et al. 2015; Arjharn et al. 2022; Prathiba et al. 2018; Das and Tiwari 2018)

| Oil characteristic | Plastic pyrolysis oil | Petroleum diesel | Gasoline |
|---|---|---|---|
| Density at 15 °C (g/cm$^3$) | 0.86–0.92 | 0.81–0.83 | 0.78 |
| Kinematic Viscosity[11] at 40 °C (mm$^2$/s) | 1.90–5.08 | 1.9–4.10 | 1.17 |
| Flash point (°C) | 30–55 | 52–78 | 42 |
| American Petroleum Institute (API)[12] Gravity at 15 °C | 24–33 | 38–39 | 55 |
| Aniline point (°C) | 30–45 | 78 | 71 |
| Pour point (°C) | −5–(−60) | 6 | – |
| Octane Number | 82–100 | – | 91–95 |
| Cetane number | 25–49 | 45–55 | – |
| High Heating Value (MJ/kg) | 41–46 | 43–46 | 43–44 |

### *3.4.2 Pyrolysis of Agricultural and Forest Residues*

Most agricultural and forest residues are composed of lignocellulosic materials which make them excellent feedstock for pyrolysis into bio-crude oil. Song et al. (2022) investigated the kinetics of pyrolysis of different kinds of forest and agricultural residues (e.g., rice straw, bagasse, coconut shell, pine waste) and concluded that when heating rates increased to 300 °C/min, the volatiles resulted in more non-condensable gases (syngas) due to the inner particle secondary reactions. The yield of hydrogen gas in the syngas generated increased from 140 mL/g at 550 °C to 279 mL/g at 850 °C (using a fixed bed reactor at a heating rate of 10 °C/min) (Song et al. 2022). In a similar report, agricultural residues such as sawdust, straw, rice husks, etc., with an optimal particle size of about 800 μm were pyrolyzed using microwave reactor (Huang et al. 2016).

### *3.4.3 Pyrolysis of Municipal Solid Waste (MSW)*

In a study to investigate the effect of pyrolysis conditions on product yields, Hasan et al. (2024) concluded that at 500–600 °C, the pyrolysis of municipal green waste of particle size 2–10 mm for 1–5 min residence time resulted in oil, biochar, and gas yields of 52.8 wt%, 21.5 wt%, and 23.7 wt%, respectively. At higher holding

[11] Kinematic viscosity is the resistance to shear or flow in a fluid, and it is a measure of a fluid's frictional property. The higher the viscosity, the lower the efficiency of its atomization (Dafsari et al. 2019).

[12] API is a measurement of the relative density of liquid fuel to water. The higher the liquid fuel density, the smaller the API.

times and temperatures around 400 °C, bio-oil and syngas yields increased whilst biochar yields decreased. Biochar production yield was investigated via pyrolysis of MSW (50 mm particle size) at 250 °C, 10 °C/min heating rate for 180 min where these optimum pyrolysis conditions resulted in about 72.62 wt% biochar yield with temperature being the most influential parameter (Saikia and Kalamdhad 2023).

The rate of non-catalytic pyrolysis of MSW was found to increase with increase in sample heating rates (20–120 °C/min, optimum heating rate < 60 °C/min), with no impact on the temperature or weight loss (Raja et al. 2023). However, zeolite-based catalytic pyrolysis of MSW was sensitive to the catalyst-to-feedstock ratio where increasing the ratio led to a decline in product yield (Raja et al. 2023).

### 3.4.4 Pyrolysis of Tire Waste

Pyrolysis is proven to effectively convert waste tires into high-quality products. Waste tire pyrolysis at 350–850 °C can generate about 33–39 wt of biochar, about 34–45 wt% of oil and about 16–33 wt% of syngas (Karagöz et al. 2020; Martínez et al. 2013). However, as with most biofuels, the amount of sulfur in pyrolysis oil depends on the type of feedstock used. Waste tire pyrolysis products contain significantly higher amount of sulfur (pyrolysis at 500–600 °C) than in standard fuel (Zhang et al. 2021) which is corrosive to equipment and detrimental to the environment during combustion. Again, the sulfur polycyclic aromatic hydrocarbon in tire waste contributes to the recalcitrant nature of the oil to be purified (Jiang et al. 2024), thus making tire pyrolysis oil unattractive compared to other waste pyrolysis oil.

Pilot-scale waste tire gasoline production via pyrolysis and the effect of modifiers to improve the quality and adhesive nature of the fuel were investigated by Pyshyev et al. (2024) with the conclusion that waste tire gasoline generated had high content of aromatics compounds with boiling points ranging from 140 to 200 °C which could not be used as fuels unless modified. Waste tire or plastic biochar obtained after pyrolysis can be useful in applications such as bitumen modifier (or asphalt binder) soot filler, sorbents, carbon black, etc., to enhance elasticity, rutting resistance, stiffness, etc. (Lee et al. 2024; Gargiulo et al. 2023).

Waste tire pyrolysis oil can effectively be blended with petroleum diesel (up to 50%) and used in automobile engines without engine modification (Karagöz et al. 2020). Murugan et al. (2008) and Doğan et al. (2012) drew similar conclusions where about 35–70% of waste tire pyrolysis oil was blended with petroleum diesel and successfully used in unmodified diesel engines. However, higher blends (>35% tire oil) have been reported to give rise to high emissions of hydrocarbons, sulfur dioxide ($SO_2$), smoke, and carbon monoxide (CO) (İlkılıç and Aydın 2011).

Catalytic pyrolysis of waste tires is found to drastically reduce the energy utilization hence reduction in the cost of production especially when low-cost eco-friendly catalysts such as red mud, fly ash, and FCC are used. The average activation energy during catalytic pyrolysis of waste tire was reduced by about 7.8–9.5 kJ/mol compared to non-catalytic pyrolysis (Wang et al. 2024). Temperature was a critical factor in determining the yield and quality of waste tire pyrolysis oil. In a fixed bed reactor on a laboratory scale, pyrolysis of waste tire at 550 °C for 90 min using 100 mL/min hydrogen as a carrier gas resulted in a high oil yield (37.25 wt%) compared to 450 °C (14.8 wt%) at same conditions (Ramirez-Canon et al. 2018). Comparatively, waste tire was subjected to pyrolysis at 450 °C in a fluidized bed reactor where tire oil yield was high (36–44 wt%) compared to fixed bed reactor (Choi et al. 2017). However, in the same conditions with auger reactor produced tire oil at a significantly lower yield (3–5 wt%) (Choi et al. 2017). Low tire oil yield (1.4–3.7 wt%) using auger pyrolysis reactor at 300 °C for 3.5 min has also been reported (Choi et al. 2016).

### 3.4.5 Pyrolysis of Sewage Sludge and Animal and Food Waste

Pyrolysis of sewage sludge is reported to improve the quality of biochar by the degradation of organic contaminants and heavy metals (Stang et al. 2022). However, the high concentrations of oxygen and nitrogen in sludge-derived pyrolysis oil lowers its stability, heating value, and energy density (Qiu et al. 2020). Nonetheless, co-pyrolysis of sewage sludge with other waste materials or biomass is found to enhance biochar quality whilst stabilizing the heavy and toxic metals in the products (Wang et al. 2021). Though there are several reports on the efficiency of co-pyrolysis of sewage sludge with microalgae, plastics, wood and agricultural residues, animal waste (Zhou et al. 2019; Ruiz-Gómez et al. 2017), etc., there are limited data on environmental impacts of the process.

Raček et al. (2024) studied the effect of catalyst (zeolite) and organic additive (wooden sawdust) on the pyrolysis of food waste into biochar at 600 °C and concluded that food waste biochar obtained via pyrolysis is an excellent source of electrical conductor due to the high salt concentration in the feedstock.

Though the high ash content (55–60%) and low calorific value (4–6 MJ/kg) of textile dyeing sludge may reduce the performance of pyrolysis, co-pyrolyzing with other waste materials especially tire and plastic waste have been reported to generate high-quality pyrolysis oil and syngas (Ding et al. 2021; Feng et al. 2022). Textile dyeing sludge which contained a significant amount of iron was pyrolyzed at gradually increasing temperatures from 150 °C to 1000 °C resulting in drastic organic degradation into high yield biochar (47.4 wt%), syngas at 43.7 wt% yield, and low oil yield at 8.9 wt% (Zhu et al. 2023). Pyrolysis of different industrial sewage sludge (paper mill sludge and textile dyeing sludge in a 1:1 ratio) with high calcium content, low ash content, and high calorific value at 600 °C led to high gas yield (39.2 wt%) compared to individual pyrolysis of paper mill sludge (24 wt% gas yield) and textile dyeing sludge (25.9 wt% gas yield) due to their strong ability to promote organic decomposition at a faster rate (Feng et al. 2022).

Pyrolysis oil obtained from food waste is found to contain high contents of oxygenated compounds and it can be purified and upgraded by co-pyrolysis with other feedstocks which have high carbon content like plastics. For instance, when waste cooking oil was co-pyrolyzed with plastics via microwave assistance (400–600 °C), high pyrolysis oil yield with minimal oxygen content, negligible amount of sulfur and nitrogen as well as high energy content of 42–46 MJ/kg were recorded (Wan Mahari et al. 2018). They further concluded that synergistic effects in achieving fast heating rates up to 50 °C/min in less than 25 min yielding about 81 wt% pyrolysis oil was demonstrated. In similar findings, Qureshi et al. (2020) also concluded that co-pyrolysis of food waste with plastic waste bottles (in a ratio of 2:3) could generate up to about 22 MJ/kg energy. Again, the temperature is reported to contribute significantly to the carbon, aromatics, and olefins yields during co-pyrolysis of waste cooking oil with other waste materials such as tea residues which led to the increase in aromatics and olefins in the oil (Wang et al. 2017). Chen et al. (2019) confirmed similar synergistic interactions between kitchen waste and tire waste in pyrolysis kinetics where the resulting oil contained high content of hydrocarbon volatiles like olefins (13–37%) and inhibition of non-hydrocarbon volatiles.

### *3.4.6 Pyrolysis of Construction and Demolishing Waste*

Construction and demolishing waste composition is discussed in Chap. 1. Pyrolysis of construction and demolition waste is found to be cost effective technology for producing high-quality gases, tar, and oil. Excavated[13] waste pyrolysis is reported to be economically feasible where the energy required to pyrolyze 1 kg of waste is about 8 MJ compared to about 29 MJ energy which can be recovered from the pyrolysis oil generated (Breyer et al. 2017). The activation energy of wood and textile excavated waste was found to be lower compared to plastics and rubber excavated waste during pyrolysis conducted at 500–900 °C (Du et al. 2022) thus allowing for a more effective and fast degradation of the waste for high yield products. Wood and rubber excavated waste generated the highest char yields of 28.5 wt% and 40.6 wt%, respectively, whilst plastic and textile excavated waste produced the highest tar yield of 74.9 wt% and highest gas yield of 61.8 wt%, respectively (Du et al. 2022). Du et al. (2022) concluded that the pyrolysis of excavated waste at high temperatures above 500 °C leads to the production of hydrogen and carbon monoxide gases but hinders $CO_2$ and methane gas formation. Liu et al. (2024) also drew a similar conclusion to Du et al. (2022), that the activation energy of excavated waste pyrolysis was greater than that of fresh waste. In their findings, plastic waste pyrolysis had high activation energy of 216–280 kJ/mol which does not favor the pyrolysis reaction for product formation. In a similar research, Kim et al. (2014) pyrolyzed construction waste wood and found out that the activation energy of the process increased gradually with temperature from 149.41 kJ/mol to 590.22 kJ/mol with complete decomposition

[13] Excavated waste, also called obsolete waste is landfill mined waste which has been degraded by microorganisms after a long landfill period (Quaghebeur et al. 2013).

of holocellulose[14] but slow degradation of lignin. At 400–500 °C, the pyrolysis of construction wood waste in a fluidized bed reactor generated high amount of oil with high phenolic compounds compared to batch reactor, signifying high degradation of lignin (Kim et al. 2014). The phenolic compounds and aromatics of the construction wood waste oil generated from the batch pyrolysis reactor were upgraded when zeolite-based catalyst (HZSM-5) was used at the same pyrolysis conditions (Kim et al. 2014).

### *3.4.7 Co-pyrolysis of Waste Materials*

Co-pyrolysis, which involves the pyrolysis of two or more different feedstock mixes, has operational cost benefits as the feedstock does not have to be sorted out or pyrolyzed separately. Again, co-pyrolysis is found to improve the quality and quantity of bio-oil and char generated (Wei et al. 2024). Co-pyrolysis of plastic waste and torrefied bamboo sawdust in a ratio of 3:1 at 30–800 °C under argon atmosphere showed very high synergism at 40 °C/min hearing rate with high hydrocarbon contents in the oil and high calorific value of the biochar (24 MJ/kg) (Alam et al. 2021). In a similar investigation with wet torrefied bamboo, soap stock solution (municipal liquid waste) was co-pyrolyzed with bamboo and results indicated that high grade oil was achieved with torrefied bamboo compared to non-treated one, implying that, in some cases depending on the kind of feedstock, pretreatment like torrefaction can improve pyrolysis oil yield and quality (Wang et al. 2018). In another report, it was concluded that co-pyrolysis of torrefied wood and plastic waste (polystyrene) resulted in a long chain of aromatics whilst the co-pyrolysis of non-treated wood with polystyrene produced phenolic derivatives and short-chain aromatics in the oil (Hassan et al. 2016).

Co-pyrolysis of a 1:1 ratio of plastic waste with lignocellulosic materials led to the reduction of coke formation and an increase in the formation of aromatics such as alkyl-benzene formed from both cellulose and olefins sources depending on the composition of plastic waste (Li et al. 2020). Co-pyrolysis of waste tire and algae biomass has both promoting and inhibiting impacts on feedstock degradation for the formation of pyrolysis oil. The ash content in waste tire (about 7–21 wt%) as well as the free radical attacking effect of algae biomass (*Spirulina sp.*) facilitates the production of pyrolysis oil whilst the inconsistency in temperature-dependent polymerization reaction (occurs at low temperatures) and atomization reaction (occurs at high temperatures) slows the release of pyrolysis oil and syngas (Xu et al. 2024).

Co-pyrolysis of a mixture of walnut shells and waste tires (3: 1 ratio with particle size of 1–3 mm) at 500 °C, 10 °C/min heating rate for 60 min on a laboratory scale resulted in about 53 wt% oil which contained about 40% light naphtha, 10% heavy naphtha, and 30% medium distillation after distillation at atmosphere, and can be

[14] Holocellulose comprises cellulose and hemicellulose.

used as gasoline or gasoline additive due to its initial boiling point (IBP) of 72 °C (Hussein et al. 2024).

Pyrolysis oil characteristics such as calorific value can be improved when different feedstocks are co-pyrolyzed. Palm shell pyrolysis oil yield (about 46.13 wt%) and calorific value (about 11.94 MJ/kg) after pyrolysis at 600 °C for 45 min. drastically increased (to about 61.63 wt% and 40.34 MJ/kg, respectively) with high aromatic and hydrocarbon content when the palm shells were co-pyrolyzed with waste tire at a 2:3 ratio under same conditions (Abnisa et al. 2013). Uçar and Karagöz (2014) also observed high aromatics and hydrocarbons contents as well as high calorific value (31–32.9 MJ/kg) for waste tire-pine nutshells (1:4 ratio) bio-oil co-pyrolyzed at 500 °C, 5 °C/min heating rate for 60 min in the presence of 50 ml/min inert gas.

A mixture of 10% medical plastic waste (predominantly syringes and plastic bottles) and 90% textile dyeing sludge was pyrolyzed at 400–550 °C at the lowest activation energy which led to the evolution of high yield methane but reduced $CO_2$ emissions (Ding et al. 2021).

### 3.4.8 *Plasma Pyrolysis of Waste Materials*

Plasma is a state of matter, a charged particle, and a conducting gas comprising ions and electrons (Frank-Kamenetskii 1972). Plasma can be thermal or non-thermal. Thermal plasma can be at high-temperature or low-temperature. High-temperature plasma is an equilibrium plasma with temperatures of $10^6$–$10^8$ K whilst low-temperature plasma ranges from 2000 to 40,000 K (Frank-Kamenetskii 1972). Ions and electrons are generated because of high power supply to a plasma carrier gas usually nitrogen, hydrogen, steam, argon, etc. For instance, for graphite plasma non-transfer arc, the electrothermal efficiency depends on the heat loss by plasma arc thus there is no requirement for a carrier gas or additional electrode cooling system (Central Pollution Control Board 2016). In thermal plasma processes, power is supplied to the electrodes which are placed in the plasma chamber where the plasma carrier gas is ionized to generate ions and electrons. Plasma torch (also called plasmatron, plasma gun, plasma arc, plasma cutter, etc.) is a device used to generate direct flow of plasma into the plasma chamber.

Plasma pyrolysis is a thermochemical conversion technology used to improve gas yield during pyrolysis of waste via the use of plasma at temperatures above 3000 K for less than one second (Bhatt et al. 2024). The plasma pyrolysis system comprises feeder, primary chamber, secondary chamber, plasma torch, burner, scrubber, cooler, pump, and chimney (Bhatt et al. 2024). During plasma pyrolysis, $N_2$ gas is injected through the Plasma arc/torch which supplies high-temperature heating (plasma jet) to generate volatile matter, converts them into light hydrocarbon gases then to syngas. Figure 3.5 shows the simplified flow diagram of plasma pyrolysis process.

At very high temperatures and in the presence of less oxygen, plasma pyrolysis cracks the components of waste faster (compared to gasification, combustion, and incineration) generating high energy particles (via ionization), combustible gas, and

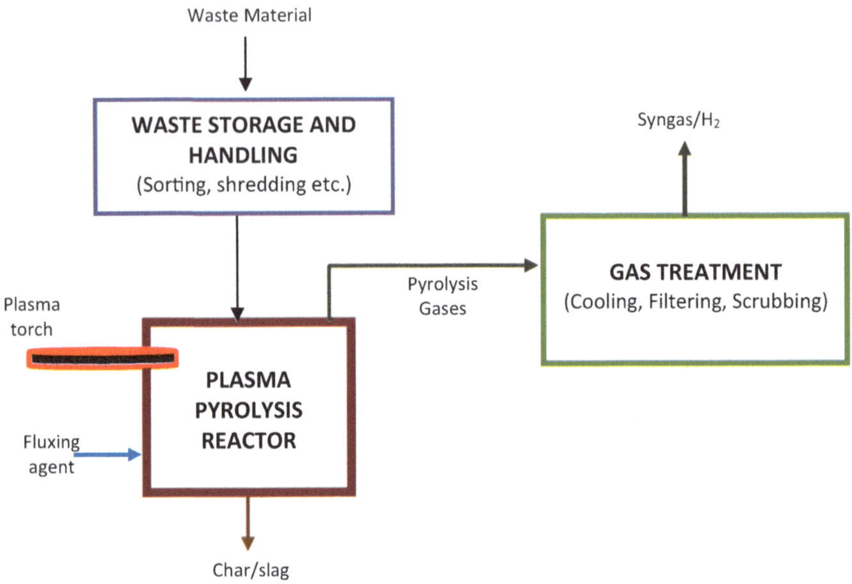


**Fig. 3.5** Simplified process flow diagram of plasma pyrolysis

solid residues. During plasma pyrolysis of waste, hydrocarbon production is reduced whilst hydrogen production is increased (Punčochář et al. 2012; Gabbar et al. 2020). Whilst conventional pyrolysis generates products with a vast variety of components, plasma pyrolysis on the other hand is an effective method which generates only gaseous and solid products with low sulfur content (Tang and Huang 2004).

Plasma pyrolysis has been used in several instances to convert waste into value-added products. Bhatt et al. (2024) used plasma pyrolysis to thermally treat plastic waste at high temperatures of 700–1000 °C where hydrogen-rich gas and biochar were produced at high yields. The generation of hydrogen during plasma pyrolysis increases with increase in arc current and temperature as reported by Mohsenian et al. (2016) who used an arc current of 80–140 A to supply heat at temperatures of 11,022–16,343 K to plasma pyrolyze plastic waste produce carbon black nano-sphere.

## 3.5 Gasification

Gasification is a thermochemical conversion technology which breaks organic matter in waste materials or biomass into syngas[15] (comprising hydrogen, carbon monoxide, carbon dioxide, etc.) and biochar via a series of endothermic and exothermic reactions at temperatures above 550 °C for 60–300 min in a controlled amount of air, steam, or oxygen (called oxidants) (Sikarwar et al. 2016; Arena 2012). During waste gasification, about 70–80% of the carbon in the waste material is converted into syngas (also called producer gas or synthetic gas) (Raheem et al. 2022) depending on various other factors like gasification temperature, residence time, gasifier type, etc. Figure 3.6 shows the flow diagram of gasification of waste into syngas for energy generation. In a gasifier, gasification of waste involves series of reactions namely:

- *Size reduction and moisture removal* of waste materials which includes grinding to increase surface area for uniform heat distribution during gasification. For efficient gasification, the biomass needs to be within a certain moisture content (about 10–40% moisture content) which is achieved by drying.
- *Pyrolysis*, which is the first step in organic matter degradation, happens in the absence of oxygen and high temperature (Bridgwater 2003). At this stage of gasification, vapors are released from the waste materials via a series of reactions, and the yield of gases generated depends on operating conditions and waste composition.
- *Oxidation*, where oxygen reacts with the waste organic matter to release CO and water vapor depending on the type of waste and oxidant used (Saghir et al. 2018). Because this stage of gasification is exothermic, heat energy is released to sustain the rest of the series of reactions following.
- *Reduction*, which is an endothermic reaction step occurring at high temperatures in the absence of oxygen. At this stage, gases are generated via a series of reactions whilst ash and char are collected as by-products (Demirbas 2006).

The syngas generated can undergo pressure swing adsorption (PSA) process to remove off-gases and clean the syngas for high-quality hydrogen gas. Again, the syngas can also undergo Fischer–Tropsch synthesis (FTS) (see Fig. 3.6) to convert it into various kinds of liquid fuels and chemicals such as diesel, methanol, etc.

During waste gasification, tar is also formed as by-product, and they can be minimized by continuous improvement and optimization of process variables such as air or steam-to-waste ratio, pressure, gasifying agents, catalyst amount, gasifier design, operating temperature, residence time, etc. (Valderrama Rios et al. 2018). Nickel catalysts are found to be highly efficient for waste gasification with the elimination of all tar formed in the process (Valderrama Rios et al. 2018). However, nickel

[15] Syngas generated from waste gasification has a lower heating value (4–40 MJ/kg) compared to fossil based natural gas (30–50 MJ/kg heating value), and consists of about 20–60 vol% CO, about 25–40 vol% $H_2$, about 5–15 vol% $CO_2$, about 0–15% $CH_4$, about 2–30 vol% $H_2O$ and other trace compounds such as $N_2$ (0.5–4 vol%), $H_2S$ (0.2–1 vol%), etc. (McKendry 2002; Couto et al. 2013; Archer and Steinberger-Wilckens 2018).

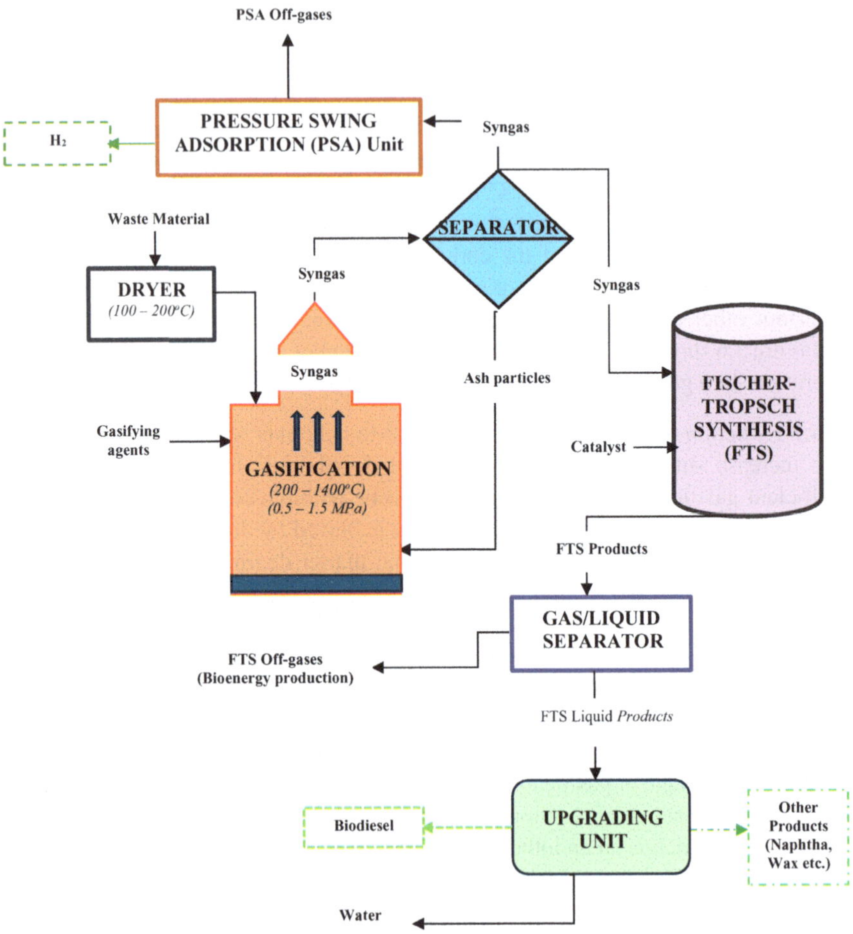


**Fig. 3.6** Simplified flow diagram of waste gasification and upgrading

catalysts get deactivated easily at a faster rate because of the deposit of carbon on its surface inhibiting its performance (Valderrama Rios et al. 2018). For instance, catalytic gasification of wood using Nickel was found to remove about 97% of tar in syngas at 800 °C reforming temperature, 15% NiO loading, and 0.3 s gas residence time (Wang et al. 2011). They concluded that though there was initial deactivation of the catalyst, it got stabilized quickly and was effective for tar removal during gasification of waste (Wang et al. 2011). Nakamura et al. (2016) obtained similar results where about 98% tar removal was recorded using bio-oil scrubber and char bed filter techniques. Because tar can condense at low temperature and clog pipes, it becomes imperative to remove as much tar as possible during gasification to make the technology more sustainable.

Gasification occurs in a gasifier (or reactor) where there is incomplete or partial oxidation to facilitate the generation of products. There are many types of gasifiers including downdraft and updraft fixed bed, fluidized bed, twin bed, entrained flow gasifiers, etc. (Arena 2012). As with all fluidized bed reactors, fluidized bed gasifiers provide uniform temperature distribution, increased temperature, easy scale-up, and easy operation (Materazzi and Lettieri 2017).

Gasification is advantageous to combustion and incineration because there is minimal amount of dioxins, furans, and $NO_x$ formation which helps improve the quality of char produced. Also, syngas generated during gasification can be upgraded into liquid fuels via Fischer–Tropsch synthesis and further used in internal combustion engines (Pan et al. 2015). Nonetheless, gasification becomes a challenge when the syngas needs to be pretreated before processing into other chemicals or fuels. Due to the use of a catalyst during gasification for improved yield, the total cost of production is increased and makes the final production costly. Again, because gasification is more suitable for wastes that are uniformly prepared, the heterogeneous nature of waste renders gasification and purification of syngas problematic (Sansaniwal et al. 2017).

Catalytic and non-catalytic gasification of waste materials has been extensively studied where gasifier designs and process conditions were optimized for high yield and quality of syngas whilst reducing tar formation. Steam gasification of waste is reported to enhance hydrogen gas concentration (as high as 30–35 mmol/g) or syngas formation from nutrient rich waste such as food waste, animal manure, etc., at very high energy recovery efficiency compared to air gasification (Valizadeh et al. 2021). In another instance, thermochemical fluidized bed gasification of rice husk was carried out on a laboratory scale by Karmakar and Datta (2011) using steam as the fluidizing/gasifying medium. They observed that steam–biomass ratio and temperature affected the hydrogen content in the product gas. On the other hand, Yan et al. (2010) who used fixed bed gasifier to convert biomass char into syngas at 600–850 °C using steam (at 0–0.357 g/min per gram of biomass) for 15 min, confirmed the results of Karmakar and Datta (2011) who concluded that steam–biomass ratio, steam flow rate (0.165 g/min/g biomass), and temperature (850 °C) influenced higher carbon conversion and syngas production.

### 3.5.1 *Plasma Gasification of Waste*

Because plasma gasification can utilize a variety of mixed waste materials for value-added products, it has become a technology of interest. One unique feature about plasma gasification is its ability to generate heat at high temperatures which helps to minimize tar formation (about 1000 times less) and increase syngas yield compared to conventional gasification, controlled incineration (waste-to-energy), and combustion (Milne et al. 1998). Again, the demand for oxygen in conventional gasification is high compared to plasma gasification since most of the heat energy is supplied from external energy source other than through exothermic reaction between the fuel and oxidant (Yoon et al. 2013). Plasma gasification uses alternating current (AC), direct current (DC) plasma torch, radio frequency, or microwaves as a heat source to decompose waste materials components into syngas (Fabry et al. 2013). Like plasma pyrolysis, the gas products from plasma gasification can be used for energy production. However, plasma gasification can produce syngas and other valuable products.

Hydrogen production yields via plasma gasification are usually lower (25–48 mol%) compared to steam gasification (27–73 mol% due to steam addition) based on operation parameters and waste composition (Mazzoni and Janajreh 2017; Gabbar et al. 2020). However, plasma pyrolysis of waste for hydrogen production via steam reforming can yield syngas concentration which is about twice (53–82.5 mol%) that of steam gasification and plasma gasification. Janajreh et al. (2013) also concluded that plasma gasification efficiency (about 42%) was lower than that of conventional gasification (about 72%).

A biomedical waste plasma gasification process was investigated by Messerle et al. (2018). In such facilities, there is a plasmatron with a rated power of 70 kW and a plasma reactor whose gasification chamber is fed with briquetted biomedical waste. During plasma gasification of biomedical waste, the air plasma torch supplies a mass-mean temperature of about 1700 K for 30 min which is able to degrade organic matter in the waste into syngas and other gases which are channeled into the cooling system whilst the slag/char is collected at the bottom of the reactor. The gas is then cleaned and used for energy production (Messerle et al. 2018).

Plasma gasification of wood waste generated high amount of hydrogen (97 g/kg wood waste) at electrical consumption of 2.23 kWh/kg compared to rice husk hydrogen (83 g/kg with 1.95 kWh/kg electricity) and sugarcane bagasse (54 g/kg with 1.48 kWh/kg electricity) (Tamayo et al. 2020). These results show lower electrical supply from the plasma torch for hydrogen generation compared to the hydrogen production using water hydrolysis process with average electrical consumption of 55 kWh/kg (Tamayo et al. 2020).

## 3.6 Torrefaction and Carbonization of Waste

Torrefaction,[16] sometimes referred to as slow or mild pyrolysis, isothermal pyrolysis, high-temperature drying and biomass baking/roasting, is a process of converting biomass/waste to biofuels at relatively low temperatures[17] (i.e., 200–390 °C for dried waste and 25–260 °C for wet waste) and pressures (2–7 MPa pressure) under inert conditions, aerobic environment or in the presence of $CO_2$ to generate biochar and

[16] Torrefaction aims to remove O, S, N, and H elemental components from the waste whilst keeping or upgrading the C to give higher heating value torrefied solid bioproduct mainly biochar (Chen et al. 2018).

[17] Torrefaction may require higher temperatures in situations where the waste materials contain lignocellulosic materials like agricultural residues, MSW, etc.).

gases via devolatilization, deoxygenation, carbonization, and depolymerization of the waste (Chen et al. 2018). Torrefaction can increase the heating value of waste materials by eliminating moisture and some volatile matter without fully degrading the organic matter content of the waste (Ighalo 2024). Torrefaction generates about 80% biochar, about 15% condensable liquid, and about 5% non-condensable gases (Acharya et al. 2012). Torrefied waste/biomass characteristics of hydrophobicity, compatibility, grindability and low microbial degradability, high energy density, low oxygen-to-carbon ratio (O/C), and easily transportable (Phusunti et al. 2018; Shankar Tumuluru et al. 2011). Torrefaction can be microwave-assisted, dry or wet (or hydrothermal carbonization), and can be carried out in similar thermochemical conversion reactors like fluidized bed reactors, moving bed reactors, etc. (Abdulyekeen et al. 2021). Process parameters such as torrefaction temperature, resident time as well as the composition of waste material or feedstock affect the yield and quality of torrefied products (Ubando et al. 2019).

During waste torrefaction, the activation energy and the degradation of carbohydrates are first altered followed by the proteins and lipids where the increment in torrefaction severity leads to increased ignition and burnout temperatures of the biochar (Chen et al. 2018). This implies that waste materials which contain high amounts of carbohydrates like food waste, are easily combustible compared to lipid-rich waste like restaurant waste.

Though carbonization is like torrefaction in many ways, carbonization can be hydrothermal where the waste is heated in water (used as catalyst or reactant) at subcritical temperatures of about 180–250 °C and 2–7 MPa pressure (Chen et al. 2018). Torrefaction below these conditions may not generate high-quality biochar. Torrefaction and carbonization of waste may be used as pretreatment options to enhance the physicochemical characteristics of the waste such as fixed carbon, moisture content, etc., for gasification, direct combustion, pyrolysis, etc. Torrefied and carbonized waste can be used as eco-friendly feedstock for electricity and heat generation via CHP plants.

## 3.7 Hydrothermal Treatment

Unlike pyrolysis, combustion, and gasification, hydrothermal treatment can efficiently convert waste materials with high water content into value-added bioproducts by eliminating the drying process which increases the operating cost of thermochemical conversion technologies.

### *3.7.1 Hydrothermal Liquefaction (HTL)*

Unlike pyrolysis which requires the feedstock to be dried, HTL can be used to process wet feedstock hence reducing energy requirement used in drying. HTL is a thermochemical conversion technology which has the capability to simultaneously treat wastewater and produce fuel from wet waste materials (Gollakota et al. 2018; Zhang and Chen 2018). HTL usually occurs at low temperatures of 200–400 °C and high pressures of 10–25 MPa with considerable amount of nitrogen in the presence or absence of catalyst, solvent (or water), and hydrogen gas where depolymerization of the wet biomass leads to the generation of liquid biofuels, biochemicals (Zhang and Chen 2018; Torri et al. 2012), and organic phase which can be treated in other biomass conversion technology systems like anaerobic digester.

During waste HTL, the polymeric units of carbohydrates, proteins, and lipids in the waste are decomposed into their monomer units which are further disintegrated into very small fragment molecules (Torri et al. 2012). At this stage, oxygen, sulfur, and nitrogen are significantly minimized whilst the hydrocarbon atoms remain in the form of low molecular weight compounds. In summary, the main sequence of reactions that occur during HTL of waste include:

1. *Depolymerization reaction* where long-chain cellulose and other polymers are broken down into short-chain hydrocarbon monomers at suitable pressure and temperature.
2. *Decomposition of short-chain monomers* by dehydration (i.e., removal of water), deamination (i.e., removal of amino acids), and decarboxylation (i.e., removal of $CO_2$). At this stage, elevated temperature and pressure usually enhance the disintegration of hydrogen bonds and efficient release of polar organic acids, phenols, furfural, etc., from the waste.
3. *Recombination and repolymerization* of smaller molecules into larger molecular weight compounds (referred to as hydrochar) due to unavailability of hydrogen.

Figure 3.7 shows the process flow diagram of hydrothermal liquefaction of waste adopted from Ofori-Boateng (2022).

The HTL bio-oil generated is upgraded into liquid biofuels like diesel (see Fig. 3.7) and high value chemicals like methanol. Like pyrolysis, HTL of waste generates high yields of liquid products but with comparatively lower oxygen and moisture content due to pressurized wet environmental conditions with HTL (Ofori-Boateng 2022). Due to the requirement of high moisture waste materials and high capital cost, the commercial scale development and adoption of HTL is still infant (Brown et al. 2010).

Catalytic HTL of wood waste (birchwood sawdust) at 300 °C for 30 min using alkaline catalysts (e.g., potassium hydroxide, potassium carbonate, colemanite, etc.) promoted the formation of aliphatic compounds and phenol derivatives via intermediate reactions (such as condensation/polymerization) and resulted in about 2–threefold higher bio-oil (about 40 wt% with high heating value HHV of 26–32 MJ/kg) and biochar yields (about 33 wt%) compared to non-catalytic HTL and acid catalyzed HTL (Nazari et al. 2015). Similarly, catalytic HTL of wheat straw at 350–400 °C and 18–35 MPa for 15 min resulted in a maximum bio-oil yield of 32.34 wt% (with heating value of 35 MJ/kg) and low biochar yield of 4.34 wt% where the oil predominantly consisted of ketones, alcohols, acids, and hydrocarbons (Seehar et al. 2020).

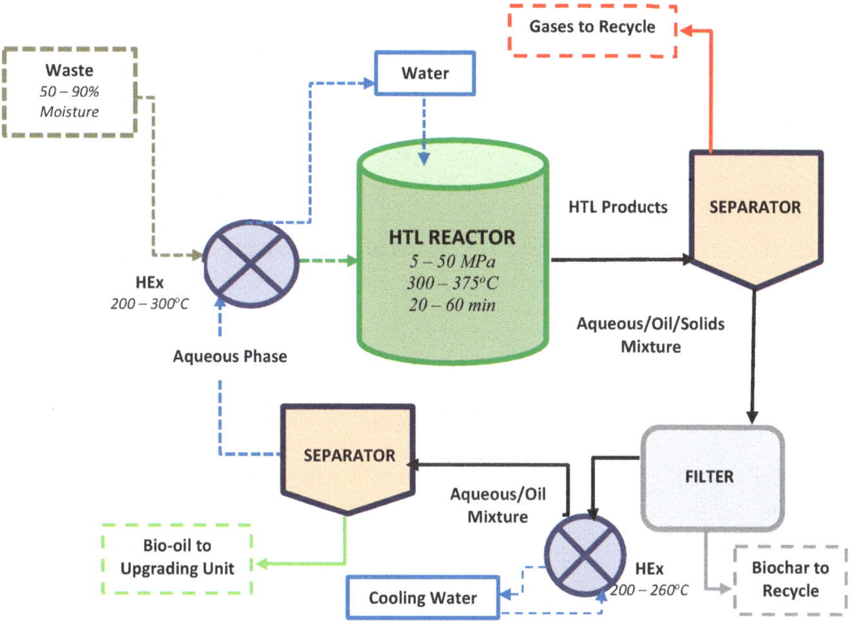


**Fig. 3.7** Simplified process flow diagram of waste hydrothermal liquefaction

### *3.7.2 Hydrothermal Gasification (HTG)*

Hydrothermal gasification (HTG) is the gasification of waste in hot compressed water, supercritical or subcritical liquid water at 400–750 °C for long residence time where high-quality syngas (e.g., hydrogen, $CO_2$, $CH_4$, CO) and hydrochar are produced (Pandey et al. 2015; Biller and Ross 2016). Water is considered supercritical above its critical point (374 °C and 22.1 MPa) at which point the vapor and liquid phases cannot be distinguished (Bröll et al. 1999). On the other hand, when water is near its critical point or below it, but with sufficient pressure above saturation pressure to preserve its liquid state, it is considered subcritical (Nanda et al. 2018). Thus, hydrothermal gasification is also referred to as either subcritical water gasification or supercritical water gasification/oxidation depending on the critical point of water considered for the process (Nanda et al. 2018; Bröll et al. 1999).

The water in the wet waste material acts as reaction medium which becomes highly reactive if heated under high pressure. At elevated temperatures above 600 °C in the presence of catalyst, $H_2$ is formed as the main product due to catalytic hydrogenation reaction (Kruse et al. 2010). Table 3.4 shows summary of HTG of waste materials.

**Table 3.4** Summary of HTG of some waste materials

| Waste material | Process conditions | | | Product yield | Reference |
|---|---|---|---|---|---|
| | Temperature, °C | Pressure, MPa | Time, min | | |
| Mix rice husks and chicken manure | 400 | 22.5 | 15 | 5.49 mol gas yield/kg | Zhang et al. (2022) |
| Mixture of glucose, molasses, and rice bran | 330–390 | 45 | 0–120 | 3–22 g $H_2$/kg waste | Muangrat et al. (2010) |
| Plastic waste and canola meal[18] | 525 | 22–25 | 60 | 8.1 mmol/g of $H_2$<br>17.9 mmol/g of total syngas | Nanda et al. (2022) |
| Soda lignin and plastic waste mix[19] | 700 | 30 | 30 | 63.3 mmol/g $H_2$ | Cao et al. (2020) |
| Waste tire | 250–600 | 22.5 | – | 41.575 mmol/g of $CH_4$ @350 °C<br>53.602 mmol/g of $H_2$ @600 °C | Nie et al. (2022) |
| Poplar waste | 300–600 | – | 60 | 20.1 mmol/g $H_2$<br>12.7 mmol/g $CH_4$ | Selvi Gökkaya et al. (2019) |
| Wheat Straw and black liquor mix[20] | 750 | 30 | 40 | 46 mmol/g $H_2$ | Cao et al. (2019) |

### *3.7.3 Hydrothermal Carbonization (HTC)*

Hydrothermal carbonization (HTC) is a thermochemical conversion technology which converts wet waste or organic matter into valuable products (mainly syngas and biochar) at very low operating temperatures (180–280 °C), moderate pressures (2–6 MPa) and moderate-to-high residence time (5–240 min) (Yoganandham et al. 2020). Water that is present in the high moisture waste materials is usually used as the solvent and reaction medium for HTC. This changes the ambiance of HTC compared to dry or conventional carbonization, and the difference in atmosphere defines the characteristics of the biochar generated. During HTC, hydrolysis takes place initially where low activation energy is exhibited with low decomposition temperature compared to dry

[18] Plastic to canola meal ratio of 0:100, 20:80, 50:50, 80:20, and 100:0 (optimum 50:50).

[19] Soda lignin to plastic ratio of 50:50 at optimum. Soda lignin added to plastic waste drastically improved hydrogen yield after HTG.

[20] Wheat straw to black liquor ratio of 1:1.

carbonization, and this reaction introduces hydroxyl and carboxyl functional groups onto the surface of the biochar produced hence improving its absorption capacity for contaminants (Libra et al. 2011).

HTC transforms the waste material into lignite-like solid biofuel called hydrochar[21] whose yield and quality are influenced by HTC process conditions. HTC usually occurs at supercritical or subcritical water conditions and at temperatures below 374 °C, the dielectric constant of water is decreased which makes water hydrogen bonds weak generating high ionization constant which facilitates the dissociation of water into acidic hydronium ions ($H_3O^+$) and basic hydroxide ions ($OH^-$) (Ischia and Fiori 2021).

High HTC temperatures of 180 and 240 °C led to reduced oxygen contents of 48.8% and 23.3% in the hydrochar, respectively, for corn straw biomass HTC (Raheem et al. 2022). Also, short HTC reaction time of 10–30 min gives rise to highest hydrochar yields (90–94.3 wt%), high calorific value (65–68 wt%) and carbon content of biochar (Raheem et al. 2022; Cai et al. 2016; Yek et al. 2022).

Table 3.5 summarizes the pros and cons of the different methods of hydrothermal treatment of waste into value-added bioproducts.

Agricultural residues such as corn stover, pine sawdust, and rapeseed straw were subjected to HTC at 170–230 °C for 3–30 min where about 51–88 wt% of hydrochar were recorded (Zhou et al. 2010). Similarly, oil palm waste (such as palm fronds, empty fruit bunches, etc.) was individually carbonized using HTC at 150–250 °C for 20–240 min with hydrochar yield of about 52–85 wt% (Zakaria et al. 2015). Forest and agricultural residue mix comprising giant bamboo, eucalyptus, coffee wood, and coffee parchment were subjected to HTC at 180–24 °C, and after 3 h of wet carbonization, the hydrochar yield recorded was about 46–56 wt% with fixed carbon content of 33–53% and high heating value of 24–30 MJ/kg (Mendoza Martinez et al. 2021). Non-catalytic HTC of sewage sludge at 140–300 °C for 30–240 min can generate high-quality hydrochar with high heating value (HHV) of about 9–32 MJ/kg depending on the reactor type and composition of sewage sludge (Zheng et al. 2019; Gaur et al. 2020; Danso-Boateng et al. 2015; Gao et al. 2019).

## 3.8 Thermochemical Hybrid Waste-to-Energy Conversion Technologies

Thermochemical conversion of waste materials to bioenergy (bioelectricity and heat) can occur in an integrated system which are renewable energy technologies (such as solar thermal, photovoltaics, fuel cells, anaerobic digestion, etc.). These integrated systems are called thermochemical hybrid waste-to-energy systems (THS), and they have evolved as one of the sustainable strategies of reducing waste and $CO_2$ emissions into the atmosphere.

---

[21] Hydrochar has improved combustion properties as solid biofuel with high hydrophobic abilities, energy density high dewaterability, and friability (Zhao et al. 2014).

**Table 3.5** Advantages and disadvantages of hydrothermal treatment of Waste

| Technology | Advantages | Disadvantages |
|---|---|---|
| Hydrothermal Carbonization[22] | • Simple and mostly occur under mild temperatures (180–280 °C)<br>• Can efficiently process wet biomass into nutrient rich biochar for soil remediation<br>• No need for drying process hence reduces energy requirement<br>• Low carbon emissions | • Low heating value biochar poses water treatment challenges<br>• Long reaction time required<br>• Fouling and slagging[23] issues due to high ash content |
| Hydrothermal Gasification[24] | • High syngas yield<br>• High process efficiency<br>• Low $NO_x$, $CO_2$, and Sox emissions<br>• Low char and tar formation<br>• Low content of nitrogen and sulfur in products | • High temperature and pressure may cause high operational costs<br>• Complex waste mix is challenging to handle due to the formation of char, metal oxides and salt precipitation<br>• Catalyst recycling issues |
| Hydrothermal Liquefaction[25] | • Tar formation reduction<br>• High product yields especially bio-oil<br>• Low-temperature operation leads to reduction in unwanted side reactions<br>• Operates at mild temperatures (200–400 °C) | • Bio-oil contains high water content which can cause corrosion of equipment<br>• Scale-up challenges with continuous processes<br>• Increased energy demand may lead to high cost of operation |

THS can increase energy storage capacities, improve energy efficiency, and power quality hence reducing energy cost compared to single-source renewable energy systems like photovoltaic (Wang et al. 2015). Also, according to Ashok (2007), THS can reduce environmental impacts, reduce electrical lines and transformer accidents, and facilitate efficient transmission and distribution of power hence rendering them reliable and safe.

[22] Cao et al. (2021), Urbanowska et al. (2020), He et al. (2022), Niinipuu et al. (2020).

[23] Slag is a sticky or molten ash and other incombustible by-products generated after combustion. Slagging occurs when slags stick to the furnace or boiler walls.

[24] Peterson et al. (2008), Okolie et al. (2021).

[25] Gollakota et al. (2018), Costa et al. (2022), Harisankar and Vinu (2023).

### 3.8.1 Incineration-Based Hybrid Waste-to-Energy Systems

In incineration-based waste power plants, the bioenergy (or electricity) generated is partially obtained from the heat produced by combustion of the waste through generators (Leme et al. 2014). During electricity generation, whilst saturated steam or hot water is produced for heat production in power plants, superheated steam is used in combined heat and power (CHP) systems.

On the other hand, for hybrid systems using integrated incineration with other renewable energy sources, clean fuel with greenhouse gas (GHG) emissions reduction makes them advantageous. For instance, an incinerated-based hybrid system with solar power uses lenses to direct large rays of sunlight onto a solar concentrator which is then transformed into heat to move a heat engine (or steam turbines). This green heat can then be channeled into the incineration facility to power it for biofuel production (Sadi and Arabkoohsar 2019). It is reported that incineration-based solar hybrid systems with capacity of 10 $MW_e$ can prevent monthly GHG emissions of about 8,000 tons of $CO_2$ equivalent compared to natural gas power plants (Sadi and Arabkoohsar 2019).

### 3.8.2 Gasification-Based Hybrid Waste-to-Energy Systems

The generation of electricity via gasification of waste is found to be environmentally benign especially when integrated into existing natural gas power plants. The configuration and design of hybrid systems as well as the composition of waste, gasifying agents (like steam, $H_2$, $O_2$, air, etc.) and turbines and generator types affect the efficiency of the hybrid system (Rokni 2015). Figure 3.8 shows the flow diagram of a gasification-based fuel cell hybrid system for power generation.

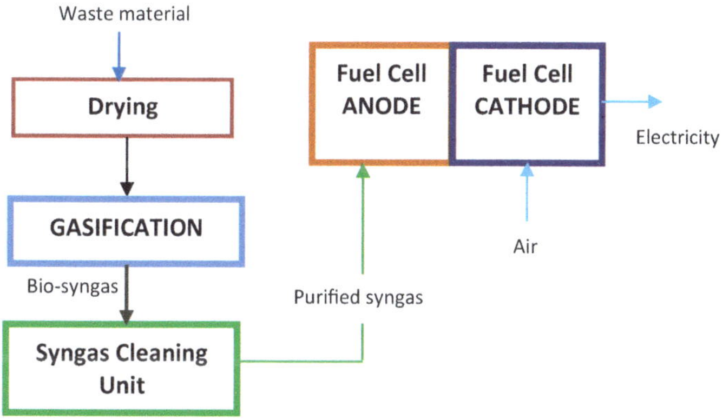


**Fig. 3.8** Process flow diagram of gasification-fuel cell hybrid system

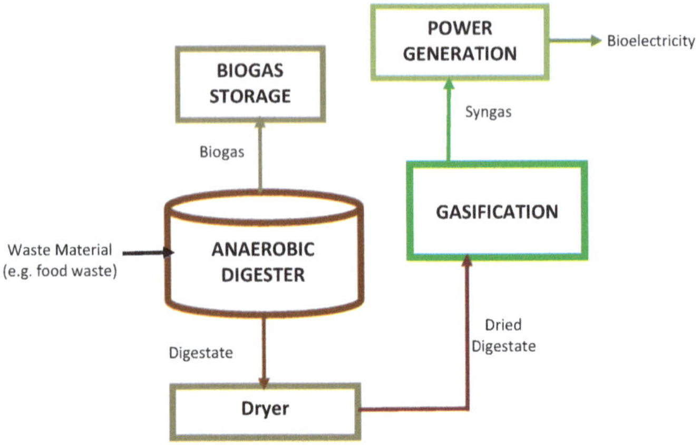


**Fig. 3.9** Process flow diagram of gasification-anaerobic digestion hybrid system

Agricultural waste gasification-based hybrid system with integrated solid oxide fuel cell[26] is reported to generate about 5.226 MW bioelectricity from a low heating value syngas at 63.6% energy efficiency with a sustainability exponent of 3.65 and 141.2 kg $CO_2$ emissions per MWh electricity (Ogorure et al. 2018). With a similar hybrid system for MSW gasification, the energy efficiency was almost the same (62.3%) as that for the agricultural waste hybrid system but with a very high net power output of 219.94 MW and high sustainability index of 6.57 (Owebor et al. 2019) perhaps due to synergistic interactions of MSW components compared to that of agricultural waste during thermal decomposition. MSW gasification-thermophotovoltaic hybrid system to produce electricity (6.77 MWh per day) and $H_2$ (2.72 tons/day) was found to be efficient (68.17%) compared to MSW gasification-molten carbonate fuel cell hybrid system whose methane gas production was about 380.1 tons/year (44.13% efficiency) (Eisavi et al. 2022; Mehrpooya et al. 2022). Figure 3.9 shows a process flow diagram of biofuel and energy production via a gasification-anaerobic digestion hybrid system.

Gasification of food waste-wood chip system was integrated with anaerobic digestion system for heat recovery with overall efficiency of about 71.3% and producer gas enthalpy of 3.2 kW (Zhang et al. 2018). In an anaerobic digestion-gasification hybrid system for cogeneration of heat and electricity from urban sewage sludge,

[26] Solid oxide fuel cells are types of fuel cells with solid oxides such as non-porous metal oxides or ceramic electrolytes which produce electricity electrochemically by oxidizing the fuel (Li et al. 2010).

Montorsi et al. (2018) reported annual production of about 810 MWh electricity and 1484.5 MWh heat which suggests a sustainable way of treating liquid waste whilst generating eco-friendly power. The efficiency and sustainability of a plasma gasification-fuel cell hybrid system for cogeneration of heat and electricity was investigated (Peng et al. 2021). In their investigation, the plasma gasifier processed 3 tons/hour of medical waste to produce syngas which was then fed into a gas turbine supported by supercritical $CO_2$ to improve the performance of the hybrid system. The CHP system recorded about 87.8% efficiency with a generation of about 14.02 MW power (about 59.3% energy efficiency). This hybrid system recycled the exhaust gases and heat from the $CO_2$ stream for hot water production which makes the system's sustainability index increase (Peng et al. 2021).

### *3.8.3 Pyrolysis-Based Hybrid Waste-to-Energy Systems*

Integrating pyrolysis technology with other renewable energy systems like solar thermal as hybrid waste-to-energy technology helps increase the overall performance and sustainability of the system. Many authors have concluded that solar thermal and photovoltaic hybrid systems integrated into pyrolysis technologies are able to effectively generate heat for sustainable biofuel and power generation (Weldekidan et al. 2018; Ghenai et al. 2020; Zeaiter et al. 2018).

With the use of solar–pyrolysis hybrid systems, changes in the weather patterns can affect the systems' efficiency due to lack of solar energy required to power the hybrid system. Zeaiter et al. (2018) saw clear distinction of energy efficiency of the hybrid system in winter (where sunshine is minimal) and summer (when sunshine is high) and concluded that, for a pyrolysis-concentrated solar power hybrid system which processes waste tire, energy savings (of the total energy requirement for the pyrolysis system alone) was about 26.6% in winter compared to about 60.8% in summer (Zeaiter et al. 2018). In a similar hybrid system for chicken litter pyrolysis, the pyrolysis reactor was supplied with heat generated from the solar thermal panel with solar radiation of about 1000 W/m$^2$ (flux density of 69,087 W/m$^2$), and as the pyrolysis temperature increased from 560 °C to 860 °C, the yield of syngas increased from 45.3 wt% to 58.6 wt% whilst pyrolysis oil mainly composed on phenols, fatty acids, nitrogen containing compounds, and sterols (Weldekidan et al. 2018).

In order to produce electricity and heat from plastic waste, a pyrolysis-photovoltaics hybrid system was employed, and the results indicated an economically viable system (cost of generated electricity was about US$ 86 per MWh) where the

GHG emissions recorded were low (162 kg $CO_2$ per MWh) compared to stand alone pyrolysis system (Ghenai et al. 2020).

Pyrolysis-anaerobic digestion hybrid systems are also reported to generate electricity and heat from waste with high energy efficiency and power output. For instance, organic waste digestate from anaerobic digestion system which was pyrolyzed at 500 °C, resulted in about 8.8 wt% syngas, 58.4 wt% bio-oil, and 32.8 wt% biochar where the syngas was used to generate electricity at 14,100 kWh per day which was about 42% higher compared to stand-alone anaerobic digestion system operating at similar conditions (Monlau et al. 2015). For a similar hybrid system but a different feedstock mix of quinoa waste and sewage sludge, the global warming potential was found to reduce by about 604 kg$CO_2$ equivalent per ton of digestate (Caiardi et al. 2022).

### *3.8.4 Hydrothermal Treatment-Based Hybrid Waste-to-Energy Systems*

HTL, HTG, and HTC technologies can be integrated into fuel cells, solar energy systems, and anaerobic digestion systems for eco-friendly and cost effective electricity production. Waste materials such as spent coffee grounds, sewage sludge, diary manure, MSW, and agricultural waste have been reported to generate high-quality products via hydrothermal treatment hybrid systems with solar thermal, photovoltaics, and anaerobic digestion.

A flow diagram of HTC-anaerobic digestion hybrid system is shown in Fig. 3.10. In Fig. 3.10, the source of biogas used for electricity generation is the biochar generated by the HTC system using the digestate from anaerobic digestion unit and other types of wet waste. HTC-anaerobic digestion hybrid systems where sewage sludge is treated for energy recovery can be operated with overall energy efficiency of about 28–70% generating electricity at 164–312.9 kWh per ton of sewage sludge (Aragón-Briceño et al. 2021; Medina-Martos et al. 2020). HTC-anaerobic digestion hybrid system for energy recovery could generate energy dense biochar with high heating

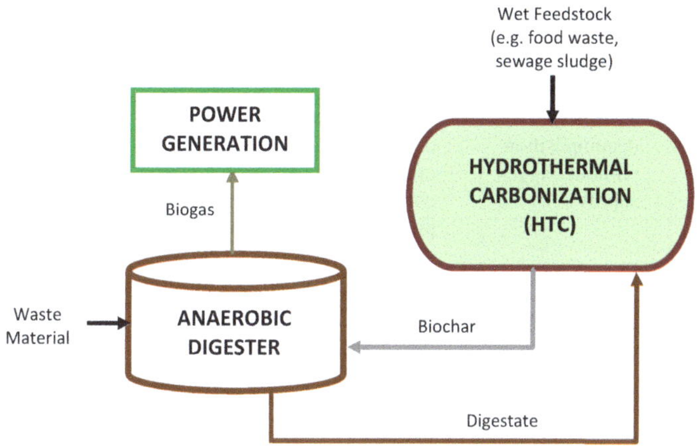


**Fig. 3.10** Process flow diagram of HTC-anaerobic digestion hybrid system

values of 24–29.6 MJ/kg agricultural residues compared to manure/whey mixture (4.4–5.3 MJ/kg) as feedstock in the same hybrid system (Reza et al. 2014; Parmar and Ross 2019; Belete et al. 2021).

HTG-solid oxide fuel cell hybrid system for power generation from organic waste with 80% moisture content was found to be sustainable with energy efficiency of 63% (Facchinetti et al. 2012). Integrating anaerobic digesters into HTL systems would facilitate energy recovery where the wet waste material is hydrothermally liquefied, and the carbon-rich aqueous phase channeled into the digester to generate biogas which can be used to produce electricity. A commercial scale HTL-anaerobic digestion hybrid waste-to-energy system utilizing about 590 million liters per day of diary manure in New York was able to generate about 607 million kWh bioelectricity (at a selling price of US$ 0.06–0.18 per kWh), 117,000 kg hydrochar, and 162,000 L bio-oil (Kassem et al. 2020). Treatment of sewage sludge in hybrid systems is more prevalent with HTC-anaerobic digestion hybrid system than HTL-anaerobic digestion hybrid system. Table 3.6 summarizes the different thermochemical conversion technologies discussed in this chapter.

**Table 3.6** Current thermochemical conversion technologies for waste treatment and biofuel and bioenergy generation

| Technology | Process summary | Advantages | Disadvantages |
|---|---|---|---|
| Pyrolysis[27] | Thermal decomposition without oxygen supply | Generation of pyrolysis oil, char, and syngas; high energy efficiency; autothermal conditions possible; improved pollution control methods; generation of diverse value-added products | High cost; toxic emissions; low energy output; generation of combustible gases require reliable control equipment |
| Combustion | Direct combustion with air or oxygen with energy recovery | Heat released is recovered for electricity generation | High cost of operation |
| Incineration[28] | Direct combustion with air or oxygen supply without energy recovery | Faster release of heat; Significant reduction of volume of waste materials; Destroys pathogens and hazardous organics; Technically and economically feasible compared to other technologies | Generation of air pollutants, dioxin, $SO_X$, $NO_X$, $CO_2$, etc Air pollution control challenges; High cost of operation; By-products (fly ash, bottom ash) are toxic |
| Gasification[29] | Thermal degradation with limited oxygen supply | Formation of producer gas, tar/char; high energy efficiency; less emissions compared to combustion and incineration; autothermal conditions possible | Generation of toxic compounds; high capital and operational cost; combustible gases may require reliable control equipment |

(continued)

[27] Abnisa and Alaba (2021), Fang et al. (2020).

[28] Malkow (2004).

[29] Das et al. (2019).

**Table 3.6** (continued)

| Technology | Process summary | Advantages | Disadvantages |
|---|---|---|---|
| Plasma Gasification[30] | High-temperature decomposition in the presence of controlled amount of oxygen | Production of syngas and slag/char; very fast heating, start-up and shut down periods; compact and requires less space; can treat different types of waste; no tar formation; reaction independent energy input; reduced gas flow | Toxic compounds may be present in gas product; high maintenance cost; high energy requirement; limited lifespan of the plasma torch with electrodes |
| Plasma Pyrolysis[31] | High-temperature degradation of waste components into gases and slag in the absence of oxygen | Dioxin, furans, etc., are with emissions limits; production of syngas, hydrogen, and slag/char; can treat all kinds of waste; very fast process; requires less space for installation and operation; no tar formation; no air/oxygen required | High energy requirement; High maintenance cost |

## 3.9 Conclusion

There exist many waste conversion technologies such as biochemical, chemical, biological, etc. processes, which are mostly feedstock composition specific. However, thermochemical conversion technologies like pyrolysis, gasification, and hydrothermal treatment are suitable for all kinds of waste including plastic waste and mixtures of municipal wastes. In this chapter, the processes involved in producing biofuels, biochemicals, and bioenergy from waste through conventional thermochemical waste conversion technologies like pyrolysis and gasification, as well as new and innovative processes like hybrid WtE technologies for energy recovery were discussed. Thermochemical waste conversion technologies were found to be efficient for bioproducts recovery. However, in order to make them more sustainable, process improvement for more energy efficient processes must be considered.

## References

Abdulyekeen KA, Umar AA, Patah MFA, Daud WMAW (2021) Torrefaction of biomass: production of enhanced solid biofuel from municipal solid waste and other types of biomass. Renew Sustain Energy Rev 150:111436

Abnisa F, Alaba PA (2021) Recovery of liquid fuel from fossil-based solid wastes via pyrolysis technique: a review. J Environ Chem Eng 9:106593

Abnisa F, Wan Daud WMA, Ramalingam S, Azemi MNBM, Sahu JN (2013) Co-pyrolysis of palm shell and polystyrene waste mixtures to synthesis liquid fuel. Fuel 108:311–318

[30] Munir et al. (2019).

[31] Central Pollution Control Board (2016), Liu et al. (2007).

Acharya B, Sule I, Dutta A (2012) A review on advances of torrefaction technologies for biomass processing. Biomass Convers Biorefin 2(4):349–369

Ahmad I, Khan MI, Khan H, Ishaq M, Tariq R, Gul K, Ahmad W (2015) Pyrolysis study of polypropylene and polyethylene into premium oil products. Int J Green Energy 12(7):663–671

Aisien E T, Otuya I C, Aisien F A (2021). Thermal and catalytic pyrolysis of waste polypropylene plastic using spent FCC catalyst. Environ Technol Innov 22:101455, 2021.

Akubo K, Nahil MA, Williams PT (2019) Aromatic fuel oils produced from the pyrolysis-catalysis of polyethylene plastic with metal-impregnated zeolite catalysts. J Energy Inst 92:195–202

Alam M, Rammohan D, Bhavanam A, Peela NR (2021) Wet torrefaction of bamboo saw dust and its co-pyrolysis with plastic. Fuel 285:119188

Al-Salem SM, Antelava A, Constantinou A, Manos G, Dutta A (2017) A review on thermal and catalytic pyrolysis of plastic solid waste (PSW). J Environ Manag 197:177–198

Aragón-Briceño CI, Ross AB, Camargo-Valero MA (2021) Mass and energy integration study of hydrothermal carbonization with anaerobic digestion of sewage sludge. Renew Energy 167:473–483

Archer SA, Steinberger-Wilckens R (2018) Systematic analysis of biomass derived fuels for fuel cells. Int J Hydrog Energy 43:23178–23192

Arena U (2012) Process and technological aspects of municipal solid waste gasification. A Rev Waste Manag 32:625–639

Arjharn W, Liplap P, Maithomklang S, Thammakul K, Chuepeng S, Sukjit E (2022) Distilled waste plastic oil as fuel for a diesel engine: fuel production, combustion characteristics, and exhaust gas emissions. ACS Omega 7:9720–9729

Ashok S (2007) Optimised model for community-based hybrid energy system. Renew Energy 32:1155–1164

Belete YZ, Mau V, Spitzer RY, Posmanik R, Jassby D, Iddya A, Kassem N, Tester JW, Gross A (2021) Hydrothermal carbonization of anaerobic digestate and manure from a dairy farm on energy recovery and the fate of nutrients. Bioresour Technol 333:125164

Bhatt KP, Patel S, Upadhyay DS, Patel RN (2024) Production of hydrogen-rich fuel gas from waste plastics using continuous plasma pyrolysis reactor. J Environ Manag 356:120446

Biller P, Ross AB (2016) Production of biofuels via hydrothermal conversion. In: Luque R, Lin CSK, Wilson K, Clark J (eds) Handbook of biofuels production, 2nd edn. Woodhead Publishing

Breyer S, Mekhitarian L, Rimez B, Haut B (2017) Production of an alternative fuel by the co-pyrolysis of landfill recovered plastic wastes and used lubrication oils. Waste Manag 60:363–374

Bridgwater AV (2003) Renewable fuels and chemicals by thermal processing of biomass. Chem Eng J 91(2):87–102

Bröll D, Kaul C, Krämer A, Krammer P, Richter T, Jung M et al (1999) Chemistry in supercritical water. Angew Chem Int Ed 38:2998–3014

Brown TM, Duan P, Savage PE (2010) Hydrothermal liquefaction and gasification of *Nannochloropsis sp.* Energy Fuels 24(6):3639–3646

Bruna Alves (2024) Plastic waste worldwide—statistics and facts, Statista. https://www.statista.com/topics/5401/global-plastic-waste/#topicOverview. Accessed 15 Apr 2024

Cahyono MS, Styana UIF (2017) Influence of heating rate and temperature on the yield and properties of pyrolysis oil obtained from waste plastic bag. Conserv J Energy Environ Stud 1:1–8

Cai J, Li B, Chen C, Wang J, Zhao M, Zhang K (2016) Hydrothermal carbonization of tobacco stalk for fuel application. Bioresour Technol 220:305–311

Cai W, Kumar R, Zheng Y, Zhu Z, Wong JWC, Zhao J (2023) Exploring the potential of clay catalysts in catalytic pyrolysis of mixed plastic waste for fuel and energy recovery. Heliyon 9(12):e23140

Caiardi F, Belaud J-P, Vialle C, Monlau F, Tayibi S, Barakat A, Oukarroum A, Zeroual Y, Sablayrolles C (2022) Waste-to-energy innovative system: assessment of integrating anaerobic digestion and pyrolysis technologies. Sustain Prod Consum 31:657–669

Cao C, Zhang Y, Li L, Wei W, Wang G, Bian C (2019). Supercritical water gasification of black liquor with wheat straw as the supplementary energy resource. Int J Hydrogen Energy 44:15737–15745.
Cao C, Bian C, Wang G, Bai B, Xie Y, Jin H (2020) Co-gasification of plastic wastes and soda lignin in supercritical water. Chem Eng J 388:124277
Cao Y, He M, Dutta S, Luo G, Zhang S, Tsang DCW (2021) Hydrothermal carbonization and liquefaction for sustainable production of hydrochar and aromatics. Renew Sust Energ Rev 152:111722
Central Pollution Control Board, Ministry of Environment and Forests and Climate change (2016) Study on plastic waste disposal through plasma pyrolysis technology. Parivesh Bhawan, East Arjun Nagar, Delhi. https://www.cpcb.nic.in/displaypdf.php?id=cGxhc3RpY3dhc3RlL1BsYXNtYS1QeXJvbHlzaXMtZmluYWwtUmVwb3J0LTIxLTExLTE2LnBkZg==. Accessed 22 Apr 2024
Chang SH (2023) Plastic waste as pyrolysis feedstock for plastic oil production: A review. Sci Total Environ 877:162719
Chen D, Gao A, Cen K, Zhang J, Cao X, Ma Z (2018) Investigation of biomass torrefaction based on three major components: hemicellulose, cellulose, and lignin. Energy Convers Manag 169:228–237
Chen J, Ma X, Yu Z, Deng T, Chen X, Chen L, Dai M (2019) A study on catalytic co-pyrolysis of kitchen waste with tire waste over ZSM-5 using TG-FTIR and Py-GC/MS. Bioresour Technol 289:121585
Cheng L, Gua J, Wang Y, Zhang J, Yuan H, Chen Y (2020) Polyethylene high-pressure pyrolysis: better product distribution and process mechanism analysis. Chem Eng J 385:123866
Choi G-G, Oh S-J, Kim J-S (2016) Scrap tire pyrolysis using a new type two-stage pyrolyzer: effects of dolomite and olivine on producing a low-sulfur pyrolysis oil. Energy 114:457–464.
Choi G-G, Oh S-J, Kim J-S (2017) Clean pyrolysis oil from a continuous two-stage pyrolysis of scrap tires using in-situ and ex-situ desulfurization. Energy 141:2234–2241
Costa PA, Mata RM, Pinto MF, Paradela F, Dutra F (2022) Hydrothermal liquefaction of microalgae for the production of biocrude and value-added chemicals. Chem Eng Trans 94:865–870
Couto N, Rouboa A, Silva V, El M, Bouziane K (2013) Influence of the biomass gasification processes on the final composition of syngas. Energy Procedia 36:596–606
Dafsari RA, Lee HJ, Han J, Park D-C, Lee J (2019) Viscosity effect on the pressure swirl atomization of an alternative aviation fuel. Fuel 240:179–191
Danso-Boateng E, Shama G, Wheatley AD, Martin SJ, Holdich RG (2015) Hydrothermal Carbonisation of Sewage Sludge: Effect of Process Conditions on Product Characteristics and Methane Production. Bioresour Technol 177:318–327
Dao HT, Djokic MR, Van Geem KM (2021) Detailed group-type characterization of plastic-waste pyrolysis oils: by comprehensive two-dimensional gas chromatography including linear, branched, and Di-olefins. Separations 8:103
Das P, Tiwari P (2018) The effect of slow pyrolysis on the conversion of packaging waste plastics (PE and PP) into fuel. Waste Manag 79:615–624
Das S, Lee SH, Kumar P, Kim KH, Lee SS, Bhattacharya SS (2019) Solid waste management: scope and the challenge of sustainability. J Clean Prod 228:658–678
Demirbas A (2006) Biomass gasification for power generation in Turkey. Energy Source Part a: Recov Utiliz Environ Effects 28(5):433–445
Di Maria F, Contini S, Bidini G, Boncompagni A, Lasagni M, Sisani F (2016) Energetic efficiency of an existing waste to energy power plant. Energy Procedia 101:1175–1182
Dincer I, Rosen MA, Khalid F (2018) Thermal energy production. In: Dincer I (ed) Comprehensive Energy Systems, vol 3. Elsevier, Energy Production, pp 673–706
Ding Z, Liu J, Chen H, Huang S, Evrendilek F, He Y, Zheng L (2021) Co-pyrolysis performances, synergistic mechanisms, and products of textile dyeing sludge and medical plastic wastes. Sci Total Environ 799:149397
Doğan O, Celik MB, Özdalyan B (2012) The effect of tire derived fuel/diesel fuel blends utilization on diesel engine performance and emissions. Fuel 95:340–346

Du Y, Ju T, Meng Y, Han S, Jiang J (2022) Pyrolysis characteristics of excavated waste and generation mechanism of gas products. J Clean Prod 370:133489
Eisavi B, Ranjbar F, Nami H, Chitsaz A (2022) Low-carbon biomass-fueled integrated system for power, methane and methanol production. Energy Convers Manag 253:115163
Environmental Protection Agency (EPA) (2024) Energy recovery from the combustion of municipal solid waste (MSW). https://www.epa.gov/smm/energy-recovery-combustion-municipal-solid-waste-msw. Accessed 12 Feb 2024
Fabry F, Rehmet C, Rohani V, Fulcheri L (2013) Waste gasification by thermal plasma: a review. Waste Biomass Valoriz 4(3):421–439
Facchinetti E, Gassner M, D'Amelio M, Marechal F, Favrat D (2012) Process integration and optimization of a solid oxide fuel cell—gas turbine hybrid cycle fueled with hydrothermally gasified waste biomass. Energy 41:408–419
Fang S, Jiang L, Li P, Bai J, Chang C (2020) Study on pyrolysis products characteristics of medical waste and fractional condensation of the pyrolysis oil. Energy 195:116969
Feng S, Zhang G, Yuan D, Li Y, Zhou Y, Lin F (2022) Co-pyrolysis of paper mill sludge and textile dyeing sludge with high calorific value solid waste: Pyrolysis kinetics, products distribution, and pollutants transformation. Fuel 329:125433
Frank-Kamenetskii DA (1972) Introduction: Plasma - Fourth State Matter. Springer, Boston, MA, pp 1–8
Gabbar HA, Lisi D, Aboughaly M, Damideh V, Hassen I (2020) Modeling of a plasma-based waste gasification system for solid waste generated onboard of typical cruiser vessels used as a feedstock. Designs 4:1–19
Gao N, Li Z, Quan C, Miskolczi N, Egedy AA (2019) New Method Combining Hydrothermal Carbonization and Mechanical Compression In-Situ for Sewage Sludge Dewatering: Bench-Scale Verification. J Anal Appl Pyrolysis 139:187–195
Gargiulo V, Alfe M, Ruoppolo G, Cammarota F, Rossi CO (2023) How char from waste pyrolysis can improve bitumen characteristics and induce anti-aging effects. Colloids Surf A: Physicochem Eng Aspects 676(Part A):132199
Gaur RZ, Khoury O, Zohar M, Poverenov E, Darzi R, Laor Y, Posmanik R (2020) Hydrothermal carbonization of sewage sludge coupled with anaerobic digestion: integrated approach for sludge management and energy recycling. Energy Convers Manag 224:113353
Ghenai C, Rasheed MA, Alshamsi MJ, Alkamali MA, Ahmad FF, Inayat A (2020) Design of hybrid solar photovoltaics/shrouded wind turbine power system for thermal pyrolysis of plastic waste. Case Stud Therm Eng 22:100773
Gollakota ARK, Kishore N, Gu S (2018) A review on hydrothermal liquefaction of biomass. Renew Sust Energ Rev 81:1378–1392
Gopalakrishnan B, Khanna N, Das D (2019) Dark-fermentative biohydrogen production. In: Mohan SV, Chang J-S, Hallenbeck PC, Larroche C (eds) Pandey A. Biohydrogen, Elsevier, pp 79–122
Günther B, Gebauer K, Barkowski R, Rosenthal M, Bues C-T (2012) Calorific value of selected wood species and wood products. Eur J Wood Prod 70:755–757
Hakeem IG, Aberuagba F, Musa U (2018) Catalytic pyrolysis of waste polypropylene using Ahoko kaolin from Nigeria. Appl Petrochem Res 8:203–210
Hariadi D, Saleh SM, Yamin RA, Aprilia S (2021) Utilization of LDPE plastic waste on the quality of pyrolysis oil as an asphalt solvent alternative. Therm Sci Eng Prog 23:100872
Harisankar S, Vinu R (2023) Comprehensive evaluation of municipal solid wastes and mixed feedstocks for commercial hydrothermal liquefaction in bio-refineries. Fuel 339:127236
Hasan MM, Rasul MG, Jahirul MI, Mofijur M (2024) Fueling the future: Unleashing energy and exergy efficiency from municipal green waste pyrolysis. Fuel 357(Part A):129815.
Hassan EB, Elsayed I, Eseyin A (2016) Production high yields of aromatic hydrocarbons through catalytic fast pyrolysis of torrefied wood and polystyrene. Fuel 174:317–324
He M, Zhu X, Dutta S, Khanal SK, Lee KT, Masek O, Tsang DCW (2022) Catalytic co-hydrothermal carbonization of food waste digestate and yard waste for energy application and nutrient recovery. Bioresour Technol 344:126395

Hu X, Gholizadeh M (2020) Progress of the applications of bio-oil. Renew Sust Energ Rev 134:110124
Huang Y-F, Chiueh P-T, Lo S-L (2016) A review on microwave pyrolysis of lignocellulosic biomass. Sustain Environ Res 26(3):103–109
Hussein AM, Kar Y, Kök OE (2024) Co-Pyrolysis of walnut shell and waste tire with a focus on bio-oil yield quantity and quality. J Energy Inst 112:101494
Ibrahim HA-H (2020) Introductory chapter: pyrolysis. In: Ibrahim HA-H (ed) Recent Advances in Pyrolysis. IntechOpen, London, pp 1–12
Ighalo JO (2024) Biowastes and derived green sorbents for water decontamination: Insights on thermochemical conversion strategies. Current Opinion in Green and Sustainable Chemistry 45:100880
İlkılıç C, Aydın H (2011) Fuel production from waste vehicle tires by catalytic pyrolysis and its application in a diesel engine. Fuel Process Technol 92(5):1129–1135
Inayat A, Inayat A, Schwieger W, Sokolova B, Lestinsky P (2022) Enhancing aromatics and olefins yields in thermo-catalytic pyrolysis of LDPE over zeolites: role of staged catalysis and acid site density of HZSM-5. Fuel 314:123071
Ischia G, Fiori L (2021) Hydrothermal carbonization of organic waste and biomass: a review on process, reactor, and plant modeling. Waste Biomass Valorization 12(6):2797–2824
Janajreh I, Raza SS, Valmundsson AS (2013) Plasma gasification process: modeling, simulation and comparison with conventional air gasification. Energy Convers Manag 65:801–809
Jegla Z, Bebar L, Pavlas M, Kropač J, Stehlik P (2010) Secondary combustion chamber with inbuilt heat transfer area—Thermal model for improved waste-to-energy systems modelling. Chem Eng Transact 21:859–864
Jiang H, Zhang J, Shao J, Fan T, Li J et al (2024) Desulfurization and upgrade of pyrolytic oil and gas during waste tires pyrolysis: the role of metal oxides. Waste Manage 182:44–54
Karagöz M, Ağbulut Ü, Sarıdemir S (2020) Waste to energy: Production of waste tire pyrolysis oil and comprehensive analysis of its usability in diesel engines. Fuel 275:117844
Karmakar MK, Datta AB (2011) Generation of hydrogen rich gas through fluidized bed gasification of biomass. Bioresour Technol 102(2):1907–1913
Kassem N, Sills D, Posmanik R, Blair C, Tester JW (2020) Combining anaerobic digestion and hydrothermal liquefaction in the conversion of dairy waste into energy: a techno economic model for New York state. Waste Manage 103:228–239
Kim JW, Lee HW, Lee I-G, Jeon J-K, Ryu C et al (2014) Influence of reaction conditions on bio-oil production from pyrolysis of construction waste wood. Renew Energy 65:41–48
Kruse A, Bernolle P, Dahmen N, Dinjus E, Maniam P (2010) Hydrothermal gasification of biomass: consecutive reactions to long-living intermediates. Energy Environ Sci 3:136–143
Law KL, Starr N, Siegler TR, Jambeck JR, Mallos NJ, Leonard GH (2020) The United States' contribution of plastic waste to land and ocean. Sci Adv 6:1–7
Lee S, Park Y-K, Lee J (2024) Upcycling of plastic and tire waste toward use as modifier for asphalt binder. Energy Environ 35(1):510–524
Leme MMV, Rocha MH, Lora EES, Venturini OJ, Lopes BM, Ferreira CH (2014) Techno-economic analysis and environmental impact assessment of energy recovery from municipal solid waste (MSW) in Brazil. Resour Conserv Recycl 87:8–20
Lerici LC, Renzini MS, Pierella LB (2015) Chemical catalyzed recycling of polymers: catalytic conversion of PE, PP and PS into fuels and chemicals over H-Y. Proc Mater Sci 8:297–303
Li M, Rao DA, Brouwer J, Samuelsen GS (2010) Design of highly efficient coal-based integrated gasification fuel cell power plants. J Power Sources 195(17):5707–5718
Li Q, Faramarzi A, Zhang S, Wang Y, Hu X, Gholizadeh M (2020) Progress in catalytic pyrolysis of municipal solid waste. Energy Convers Manag 226:113525
Liao W, Zhang X, Fu Z, Zhang S, Shao J, Yang H, Wang X, Chen H (2023) Emission characteristics of typical gas pollutants during oxygen-enriched waste incineration process. Environ Technol Innov 32:103358

Libra JA, Ro KS, Kammann C, Funke A, Berge ND, Neubauer Y, Titirici M-M, Fühner C, Bens O, Kern J, Emmerich K-H (2011) Hydrothermal carbonization of biomass residuals: a comparative review of the chemistry, processes and applications of wet and dry pyrolysis. Biofuels 2(1):71–106

Liu B, Kikuchi M, Li H, Iwao T, Inaba T (2007) Dual torch plasma arc furnace for medical waste treatment. Plasma Sci Technol 9(6):709–712

Liu Y, Tao J, Li J, Li H, Li F, Cheng Z, Yan B, Chen G (2024) Pyrolysis of combustible fractions in excavated waste: Effect of landfill time on pyrolysis characteristics analyzed by TG-FTIR-MS. J Anal Appl Pyrolysis 177:106298

López A, de Marco I, Caballero BM, Laresgoiti MF, Adrados A (2011) Influence of time and temperature on pyrolysis of plastic wastes in a semi-batch reactor. Chem Eng J 173:62–71

Ma C, Li B, Chen D, Wenga T, Ma W, Lin F, Chen G (2019) An investigation of an oxygen-enriched combustion of municipal solid waste on flue gas emission and combustion performance at a 8 MWth waste-to-energy plant. Waste Manag 96:47–56

Malkow T (2004) Novel and innovative pyrolysis and gasification technologies for energy efficient and environmentally sound MSW disposal. Waste Manag 24:53–79

Marathe PS, Westerhof RJM, Kersten SRA (2020) Effect of pressure and hot vapor residence time on the fast pyrolysis of biomass: experiments and modeling. Energy Fuel 34:1773–1780

Martínez JD, Puy N, Murillo R, García T, Navarro MV, Mastral AM (2013) Waste tyre pyrolysis—a review. Renew Sustain Energy Rev 23:179–213

Materazzi M, Lettieri P (2017) Fluidized beds for the thermochemical processing of waste. Reference Module in Chemistry Molecular Sciences and Chemical Engineering. https://doi.org/10.1016/B978-0-12-409547-2.12180-8

Mazzoni L, Janajreh I (2017) Plasma gasification of municipal solid waste with variable content of plastic solid waste for enhanced energy recovery. In: Proceedings of the International Renewable and Sustainable Energy Conference IRSEC, pp 907–912

McKendry P (2002) Energy production from biomass (part 3): Gasification technologies. Bioresour Technol 83(1):55–63

Medina-Martos E, Istrate I-R, Villamil JA, Gálvez-Martos J-L, Dufour J, Mohedano ÁF (2020) Techno-economic and life cycle assessment of an integrated hydrothermal carbonization system for sewage sludge. J Clean Prod 277:122930

Mehrpooya M, Ghorbani A, Ali Moosavian SM, Amirhaeri Y (2022) Optimal design and economic analysis of a hybrid process of municipal solid waste plasma gasification, thermophotovoltaic power generation and hydrogen/liquid fuel production. Sustain Energy Technol Assess 49:101717

Mendoza Martinez CL, Sermyagina E, Saari J, Silva de Jesus M, Cardoso M, Matheus de Almeida G, Vakkilainen E (2021) Hydrothermal Carbonization of Lignocellulosic Agro-Forest Based Biomass Residues. Biomass Bioenergy 147:106004

Messerle VE, Mosse AL, Ustimenko AB (2018) Processing of biomedical waste in plasma gasifier. Waste Manag 79:791–799

Milne TA, Abatzoglou N, Evans RJ (1998) Biomass gasifier "Tars": their nature, formation, and conversion, vol 570. National Renewable Energy Laboratory, Golden, CO. https://www.nrel.gov/docs/fy99osti/25357.pdf. Accessed 17 Feb 2024

Mohsenian S, Esmaili MS, Fathi J, Shokri B (2016) Hydrogen and carbon black nano-spheres production via thermal plasma pyrolysis of polymers. Int J Hydrog Energy 41:16656–16663

Monlau F, Sambusiti C, Antoniou N, Barakat A, Zabaniotou A (2015) A new concept for enhancing energy recovery from agricultural residues by coupling anaerobic digestion and pyrolysis process. Appl Energy 148:32–38

Montorsi L, Milani M, Venturelli M (2018) Economic assessment of an integrated waste to energy system for an urban sewage treatment plant: a numerical approach. Energy 158:105–110

Moško J, Pohořelý M, Zach B, Svoboda K, Durda T, Jeremiáš M, Šyc M, Václavková Š, Skoblia S, Beňo Z, Brynda J (2018) Fluidized bed incineration of sewage sludge in $O_2/N_2$ and $O_2/CO_2$ atmospheres. Energy Fuels 32(2):2355–2365

Muangrat R, Onwudili JA, Williams PT (2010) Alkali-promoted hydrothermal gasification of biomass food processing waste: a parametric study. Int J Hydrog Energy 35:7405–7415
Munir MT, Mardon I, Al-Zuhair S, Shawabkeh A, Saqib NU (2019) Plasma gasification of municipal solid waste for waste-to-value processing. Renew Sustain Energy Rev 116:109461
Murata K, Sato K, Sakata Y (2004) Effect of pressure on thermal degradation of polyethylene. J Anal Appl Pyrolysis 71:569–589
Murugan S, Ramaswamy MC, Nagarajan G (2008) The use of tyre pyrolysis oil in diesel engines. Waste Manage 28(12):2743–2749
Nakamura S, Kitano S, Yoshikawa K (2016) Biomass gasification process with the tar removal technologies utilizing bio-oil scrubber and char bed. Appl Energy 170:186–192
Nanda S, Reddy SN, Vo D-VN, Sahoo BN, Kozinski JA (2018) Catalytic gasification of wheat straw in hot compressed (subcritical and supercritical) water for hydrogen production. Energy Sci Eng 6:448–459
Nanda S, Okolie J A, Patel R, Pattnaik F, Fang Z, Dalai A K, Kozinski J A, Naik S (2022). Catalytic hydrothermal co-gasification of canola meal and low-density polyethylene using mixed metal oxides for hydrogen production. Int J Hydrogen Energy 47(100):42084–42098.
National Research Council (2000) Incineration Processes and Environmental Releases: Waste Incineration and Public Health. The National Academies Press Washington, DC
Nazari L, Yuan Z, Souzanchi S, Ray MB, Xu C (Charles) (2015) Hydrothermal liquefaction of woody biomass in hot-compressed water: catalyst screening and comprehensive characterization of bio-crude oils. Fuel 162:74–83
Nie SQ, Chen MQ, Li QH (2022) Evaluation on hydrothermal gasification of waste tires based on chemical equilibrium analysis. Int J Hydrog Energy 47:1435–1448
Niinipuu M, Latham KG, Boily J-F, Bergknut M, Jansson S (2020) The impact of hydrothermal carbonization on the surface functionalities of wet waste materials for water treatment applications. Environ Sci Pollut Res 27:24369–24379
Ofori-Boateng C (2022) Algal Biorefining: Resource Expenditure and Exergo-environmental Sustainability. Nova Science Publishers, New York, p 296
Ogorure OJ, Oko COC, Diemuodeke EO, Owebor K (2018) Energy, exergy, environmental and economic analysis of an agricultural waste-to-energy integrated multigeneration thermal power plant. Energy Convers Manag 171:222–240
Okolie JA, Nanda S, Dalai AK, Kozinski JA (2021) Techno-economic evaluation and sensitivity analysis of a conceptual design for supercritical water gasification of soybean straw to produce hydrogen. Bioresour Technol 331:125005
Organization for Economic Co-operation and Development (OECD) (2022) Global plastic outlook: Economic drivers, environmental impacts and policy options, OECD Publishing, Paris. https://doi.org/10.1787/de747aef-en. Accessed 11 Apr 2024
Owebor K, Oko COC, Diemuodeke EO, Ogorure OJ (2019) Thermo-environmental and economic analysis of an integrated municipal waste-to-energy solid oxide fuel cell, gas-, steam-, organic fluid- and absorption refrigeration cycle thermal power plants. Appl Energy 239:1385–1401
Palacio JCE, Santos José JCS, Renó MLG, Júnior JCF, et al (2018) Municipal Solid Waste Management and Energy Recovery. In: Al-Bahadly IH (ed) Energy Conversion-current technologies and future trends, chapter 8. IntechOpen, Rijeka
Palanivelrajan AR, Feroskhan M (2023) Comparison of various catalysts in pyrolysis process: A review. Mater Today Proc 84:47–56
Pan S-Y, Du MA, Huang IT, Liu IH, Chang EE, Chiang P-C (2015) Strategies on implementation of waste-to-energy (WTE) supply chain for circular economy system: a review. J Clean Prod 108:409–421
Pandey A, Bhaskar T, Stöcker M, Sukumaran RK (eds) (2015) Recent advances in thermo-chemical conversion of biomass. Elsevier, Boston, pp 251–267
Park K-B, Choi M-J, Chae D-Y, Jung J, Kim J-S (2022) Separate two-step and continuous two-stage pyrolysis of a waste plastic mixture to produce a chlorine-depleted oil. Energy 244:122583

Parmar KR, Ross AB (2019) Integration of hydrothermal carbonisation with anaerobic digestion; opportunities for valorisation of digestate. Energies 12:1586

Peng W, Chen H, Liu J, Zhao X, Xu G (2021) Techno-economic assessment of a conceptual waste-to-energy CHP system combining plasma gasification, SOFC, gas turbine and supercritical $CO_2$ cycle. Energy Convers Manag 245:114622

Peng Y, Wang Y, Ke L, Dai L, Wu Q, Cobb K, Zeng Y, Zou R, Liu Y, Ruan R (2022) A review on catalytic pyrolysis of plastic wastes to high-value products. Energy Convers Manag 254:115243

Peterson AA, Vogel F, Lachance RP, Fröling M, Antal JMJ, Tester JW (2008) Thermochemical biofuel production in hydrothermal media: A review of sub- and supercritical water technologies. Energy Environ Sci 1:32–65

Phusunti N, Phetwarotai W, Tekasakul S (2018) Effects of torrefaction on physical properties, chemical composition and reactivity of microalgae. Korean J Chem Eng 35(2):503–510

Prathiba R, Shruthi M, Miranda LR (2018) Pyrolysis of polystyrene waste in the presence of activated carbon in conventional and microwave heating using modified thermocouple. Waste Manag 76:528–536

Prurapark R, Owjaraen K, Saengphrom B, Limthongtip I (2020) Effect of temperature on pyrolysis oil using high density polyethylene and polyethylene terephthalate sources from mobile pyrolysis plant. Front Energy Res 8. https://www.frontiersin.org/articles/10.3389/fenrg.2020.541535/full

Punčochář M, Ruj B, Chatterjee PK (2012) Development of process for disposal of plastic waste using plasma pyrolysis technology and option for energy recovery. Procedia Eng 42:420–430

Pyshyev S, Lypko Y, Korchak B, Poliuzhyn I, Hubrii Z, Pochapska I, Rudnieva K (2024) Study on the composition of gasoline fractions obtained as a result of waste tires pyrolysis and production bitumen modifiers from it. J Energy Institute 114:101598

Qiu Z, Zhai Y, Li S, Liu X, Liu X, Wang B et al (2020) Catalytic co-pyrolysis of sewage sludge and rice husk over biochar catalyst: bio-oil upgrading and catalytic mechanism. Waste Manag 114:225–233

Quaghebeur M, Laenen B, Geysen D, Nielsen P, Pontikes Y, Gerven TV, Spooren J (2013) Characterization of landfilled materials: screening of the enhanced landfill mining potential. J Clean Prod 55:72–83

Qureshi U, Imtiaz B, Jamal Y (2020) Synthesizing PET and food waste into refuse plastic fuel (RPF): optimization and kinetic modeling. J Therm Anal Calorim 140:1745–1758

Raček J, Chorazy T, Miino MC, Vršanská M, Brtnický M, Mravcová L, Kučerík J, Hlavínek P (2024) Biochar production from the pyrolysis of food waste: Characterization and implications for its use. Sustain Chem Pharm 37:101387

Raheem A, Ding L, He Q, Mangi HF, Khand ZH, Sajid M, Ryzhkov A, Yu G (2022) Effective pretreatment of corn straw biomass using hydrothermal carbonization for co-gasification with coal: response surface methodology–Box Behnken design. Fuel 324:124544

Raja N, Monsalve-Bravo GM, Kaneti YV, Mensah J, Wilson K, Lee AF, Konarova M (2023) Thermogravimetric kinetic analysis of catalytic and non-catalytic pyrolysis of simulated municipal solid waste. Chem Eng J 470:144046

Ramirez-Canon A, Muñoz-Camelo YF, Singh P (2018) Decomposition of used Tyre Rubber by pyrolysis: enhancement of the physical properties of the liquid fraction using a hydrogen stream. Environ 5(6):72

Reza MT, Werner M, Pohl M, Mumme J (2014) Evaluation of integrated anaerobic digestion and hydrothermal carbonization for bioenergy production. J vis Exp 88:e51734

Riesco-Avila JM, Vera-Rozo JR, Rodríguez-Valderrama DA, Pardo-Cely DM, Ramón-Valencia B (2020) Effects of heating rate and temperature on the yield of thermal pyrolysis of a random waste plastic mixture. Sustainability 14(15):9026

Rokni M (2015) Thermodynamic analyses of municipal solid waste gasification plant integrated with solid oxide fuel cell and Stirling hybrid system. Int J Hydrog Energy 40:7855–7869

Ruiz-Gómez N, Quispe V, Ábrego J, Atienza-Martínez M, Murillo MB, Gea G (2017) Co-pyrolysis of sewage sludge and manure. Waste Manag 59:211–221

Sadi M, Arabkoohsar A (2019) Modelling and analysis of a hybrid solar concentrating-waste incineration power plant. J Clean Prod 216:570–584

Saghir M, Rehan M, Nizami A-S (2018) Recent Trends in Gasification Based Waste-to-Energy. In: Yun Y (ed) Gasification for low-grade feedstock. InTech. https://doi.org/10.5772/intechopen.74487

Saikia S, Kalamdhad AS (2023) Response surface methodology (RSM)-based pyrolysis process parameter optimization for char generation from municipal solid waste (MSW) in a fixed bed reactor. Sustain Energy Technol Assess 60:103541

Sansaniwal SK, Rosen MA, Tyagi SK (2017) Global challenges in the sustainable development of biomass gasification: an overview. Renew Sustain Energy Rev 80:23–43

Seehar TH, Toor SS, Shah AA, Pedersen TH, Rosendahl LA (2020) Biocrude Production from Wheat Straw at Sub and Supercritical Hydrothermal Liquefaction. Energies 13:3114

Sekar M, Ponnusamy VK, Pugazhendhi A, Praveenkumar TR, Nizeti S (2022) Production and utilization of pyrolysis oil from solid plastic wastes: a review on pyrolysis process and influence of reactors design. J Environ Manag 302:114046

Selvi Gökkaya D, Çokkuvvetli T, Sağlam M, Yüksel M, Ballice L (2019) Hydrothermal gasification of poplar wood chips with alkali, mineral, and metal impregnated activated carbon catalysts. J Supercrit Fluids 152:104542

Shankar Tumuluru J, Sokhansanj S, Hess JR, Wright CT, Boardman RD (2011) A review on biomass torrefaction process and product properties for energy applications. Ind Biotechnol 7(5):384–401

Sikarwar VS, Zhao M, Clough P, Yao J, Zhong X, Memon MZ et al (2016) An overview of advances in biomass gasification. Energy Environ Sci 9:2939–2977

Singh RK, Ruj B, Sadhukhan AK, Gupta P (2019) Thermal degradation of waste plastics under non-sweeping atmosphere: part 1: effect of temperature, product optimization, and degradation mechanism. J Environ Manag 239:395–406

Singh TS, Verma TN, Singh HN (2020) A lab scale waste to energy conversion study for pyrolysis of plastic with and without catalyst: engine emissions testing study. Fuel 277:118176

Sonawane YB, Shindikar MR, Khaladkar MY (2017) High calorific value fuel from household plastic waste by catalytic pyrolysis. Nat Environ Pollut Technol 16:879–882

Song G, Huang D, Li H, Wang X, Ren Q et al (2022) Pyrolysis reaction mechanism of typical Chinese agriculture and forest waste pellets at high heating rates based on the photo-thermal TGA. Energy 244(Part B):123164

Soni VK, Singh G, Vijayan BK, Chopra A, Kapur GS, Ramakumar SSV (2021) Thermochemical recycling of waste plastics by pyrolysis: a review. Energy Fuel 35:12763–12808

Stang C, Mohamed BA, Li LY (2022) Microplastic removal from urban stormwater: current treatments and research gaps. J Environ Manag 317:115510

Syamsiro M, Saptoadi H, Norsujianto T, Noviasri CS, Alimuddin Z, Yoshikawa K (2014) Fuel oil production from Municipal plastic wastes in sequential pyrolysis and catalytic reforming reactors. Energy Proc 47:180–188

Tamayo J, Peña L, Vázquez A, Brito Á (2020) Hydrogen-Rich Syngas Production by Plasma Gasification of Existing Biomasses in Cuba. Rev Cienc Técnicas Agropecu 29:53–63

Tang L, Huang H (2004) An investigation of sulfur distribution during thermal plasma pyrolysis of used tires. J Anal Appl Pyrolysis 72:35–40

Torri C, Garcia Alba L, Samorì C, Fabbri D, Brilman DWFW (2012) Hydrothermal treatment (HTT) of microalgae: detailed molecular characterization of HTT oil in view of HTT mechanism elucidation. Energy Fuels 26:658–671

Ubando AT, Rivera DRT, Chen W-H, Culaba AB (2019) A comprehensive review of life cycle assessment (LCA) of microalgal and lignocellulosic bioenergy products from thermochemical processes. Bioresour Technol 291:121837

Uçar S, Karagöz S (2014) Co-pyrolysis of pine nut shells with scrap tires. Fuel 137:85–93

Urbanowska A, Kabsch-Korbutowicz M, Wnukowski M, Seruga P, Baranowski M et al (2020) Treatment of liquid by-products of hydrothermal carbonization (HTC) of agricultural digestate using membrane separation. Energies 13(1):262

Valderrama Rios ML, González AM, Lora EES, Almazán del Olmo OA (2018) Reduction of tar generated during biomass gasification: a review. Biomass Bioenergy 108:345–370
Valizadeh S, Lam SS, Ko CH, Lee SH, Farooq A, Yu YJ et al (2021) Biohydrogen production from catalytic conversion of food waste via steam and air gasification using eggshell- and homo-type Ni/Al2O3 catalysts. Bioresour Technol 320:124313
Wan Mahari WA, Chong CT, Cheng CK, Lee CL, Hendrata K, Yuh PNY, Ma NL, Lam SS (2018) Production of value-added liquid fuel via microwave co-pyrolysis of used frying oil and plastic waste. Energy 162:309–317
Wang D, Yuan W, Ji W (2011) Char and char-supported nickel catalysts for secondary syngas cleanup and conditioning. Appl Energy 88:1656–1663
Wang J, Zhong Z, Zhang B, Ding K, Xue Z, Deng A, Ruan R (2017) Upgraded bio-oil production via catalytic fast co-pyrolysis of waste cooking oil and tea residual. Waste Manage 60:357–362
Wang X, Palazoglu A, El-Farra NH (2015) Operational optimization and demand response of hybrid renewable energy systems. Appl Energy 143:324–335
Wang Y, Wu Q, Dai L, Zeng Z, Liu Y, Ruan R, Fu G, Yu Z, Jiang L (2018) Co-pyrolysis of wet torrefied bamboo sawdust and soapstock. J Anal Appl Pyrolysis 132:211–216
Wang X, Chang VWC, Li Z, Chen Z, Wang Y (2021) Co-pyrolysis of sewage sludge and organic fractions of municipal solid waste: synergistic effects on biochar properties and the environmental risk of heavy metals. J Hazard Mater 412:125200
Wang K, Shan T, Li B, Zheng Y, Xu H, Wang C, Tian X (2024) Study on pyrolysis characteristics, kinetics and thermodynamics of waste tires catalytic pyrolysis with low-cost catalysts. Fuel 356:129644
Wei X, Liu P, Huang S, Wu Y, Wu S (2024) A comprehensive study on product properties and synergistic effects of co-pyrolysis of antibiotic mycelium residues and agricultural and forestry wastes. Fuel 363:131015
Weldekidan H, Strezov V, Kan T, Town G (2018) Waste to energy conversion of chicken litter through a solar-driven pyrolysis process. Energy Fuels 32:4341–4349
Wen Z, Xie Y, Chen M, Dinga CD (2021) China's plastic import ban increases prospects of environmental impact mitigation of plastic waste trade flow worldwide. Nat Commun 12:425
Williams P T (2013) Waste treatment and disposal. John Wiley Publishers
Xu Q, Xian S, Su Y, Li H, Wu Y (2024) The promoting and inhibitory mechanism of interaction between waste tires and Spirulina on their co-pyrolysis. Fuel 368:131606
Yan F, Luo SY, Hu ZQ, Xiao B, Cheng G (2010) Hydrogen-rich gas production by steam gasification of char from biomass fast pyrolysis in a fixed-bed reactor: influence of temperature and steam on hydrogen yield and syngas composition. Bioresour Technol 101(14):5633–5637
Yek PNY, Liew RK, Wan WA et al (2022) Production of value-added hydrochar from single-mode microwave hydrothermal carbonization of oil palm waste for de-chlorination of domestic water. Sci Total Environ 833:154968
Yoganandham ST, Sathyamoorthy G, Renuka RR (2020) Emerging extraction techniques: hydrothermal processing. In: Kraan S, Dominguez H (eds) Torres MD. Sustainable Seaweed Technologies, Elsevier, pp 191–205
Yoon SJ, Yun YM, Seo MW, Kim YK, Ra HW, Lee J-G (2013) Hydrogen and syngas production from glycerol through microwave plasma gasification. Int J Hydrog Energy 38(34):14559–14567
Zakaria MR, Hirata S, Fujimoto S, Hassan MA (2015) Combined pretreatment with hot compressed water and wet disk milling opened up oil palm biomass structure resulting in enhanced enzymatic digestibility. Bioresour Technol 193:128–134
Zaman AU (2010) Comparative study of municipal solid waste treatment technologies using life cycle assessment method. Int J Environ Sci Technol 7:225–234
Zanella D, Romagnoli M, Malcangi S, Beccaria M, Chenet T, De Luca C et al (2023) The contribution of high-resolution GC separations in plastic recycling research. Anal Bioanal Chem 415:2343–2355
Zeaiter J, Azizi F, Lameh M, Milani D, Ismail HY, Abbas A (2018) Waste tire pyrolysis using thermal solar energy: an integrated approach. Renew Energy 123:44–51

Zhang L. (2021). China: Single-use plastic straw and bag ban takes effects. Libr. Congr. https://www.loc.gov/item/global-legalmonitor/2021-03-23/china-single-use-plastic-straw-and-bag-ban-takeseffect/#:~:text=Article%20China%3A%20Single%2DUse%20Plastic%20Straw%20and%20Bag%20Ban%20Takes%20Effect&text=(Mar.,from%20providing%20plastic%20shopping%20bags.

Zhang G, Feng C, Zhang Y, Zhao L, Jingye C, Liyuan C, Gao J, Xu C (2021) Properties and utilization of waste tire pyroysis oil: a mini review. Fuel Process Technol 211:106582

Zhang H, Zhang R, Li W, Ling Z, Shu W, Ma J, Yan Y (2022) Agricultural waste-derived biochars from co-hydrothermal gasification of rice husk and chicken manure and their adsorption performance for dimethoate. J Hazard Mater 429:128248

Zhang J, Kan X, Shen Y, Loh K-C, Wang C-H, Dai Y, Tong YW (2018) A hybrid biological and thermal waste-to-energy system with heat energy recovery and utilization for solid organic waste treatment. Energy 152:214–222

Zhang L (2017) China: National plan on banning foreign garbage and reducing solid waste imports. Library of Congress. https://www.loc.gov/item/global-legal-monitor/2017-08-08/china-national-plan-on-banning-foreign-garbage-and-reducing-solid-waste-imports/. Accessed 23 Apr 2024

Zhang Y, Chen WT (2018) Hydrothermal liquefaction of protein-containing feedstocks. In: Thermochemical D (ed) Rosendahl L. Woodhead Publishing, Liquefaction for Energy Applications, pp 127–168

Zhao P, Shen Y, Ge S, Chen Z, Yoshikawa K (2014) Clean solid biofuel production from high moisture content waste biomass employing hydrothermal treatment. Appl Energy 131:345–367

Zheng C, Ma X, Yao Z, Chen X (2019) The Properties and Combustion Behaviors of Hydrochars Derived from Co-Hydrothermal Carbonization of Sewage Sludge and Food Waste. Bioresour Technol 285:121347

Zhou YLY, Wan C, Li D, Mao Z (2010) Effect of hot water pretreatment severity on the degradation and enzymatic hydrolysis of corn stover. Trans ASABE 53:1929–1934

Zhou J, Qiao Y, Wang W, Leng E, Huang J, Yu Y et al (2016) Formation of styrene monomer, dimer and trimer in the primary volatiles produced from polystyrene pyrolysis in a wire-mesh reactor. Fuel 182:333–339

Zhou Y, Liu Y, Jiang W, Shao L, Zhang L, Feng L (2019) Effects of pyrolysis temperature and addition proportions of corncob on the distribution of products and potential energy recovery during the preparation of sludge activated carbon. Chemosphere 221:175–183

Zhu J, Lin S, Fan Y, Chen Y, Jin L, Gao H (2023) Investigations on the iron-rich textile dyeing sludge pyrolysis characteristics: Thermal decomposition behaviors, products, potential mechanisms. J Anal Appl Pyrolysis 169:105834

# Chapter 4
# Recent Development, Challenges, and Breakthroughs of Thermochemical Conversion Technologies

**Abstract** An extensively explored biofuel alternative technology with many advantages and high conversion efficiency is thermochemical conversion through which about 30–70% of biofuels and bioenergy in the world are obtained, rendering them one of the sustainable biofuel, bioenergy, and biochemicals technologies. Nonetheless, thermodynamic analysis, technoeconomic assessment as well as environmental analyses have revealed thermochemical conversion technologies as complicated, energy-intensive, and costly. There are many challenges faced by thermochemical conversion facilities in the world though they are potentially sustainable candidates for eco-friendly fuels and chemicals in the long term. This chapter dives into the main sustainability challenges and solutions, current technological development, and prospects of thermochemical conversion technologies for biofuels, bioenergy, and biochemicals production.

**Keywords** Waste · Bioenergy · Biofuel · Thermochemical conversion prospects · Ultrasonic-assisted pyrolysis · Hybrid WtE · Gasification · Pyrolysis

## 4.1 Introduction

The advancement in information technology, sustainability, research, and development (R&D) are highly embraced by the world to improve standard of living. However, waste management has been a global concern as population increases daily and waste generation rate rises. Coupled with environmental concerns of fossil fuels (which form about 85% of the world's primary energy use), biofuels have become potential alternatives, and extensive R&D are ongoing for continuous improvement in bioenergy and biofuel technologies on both commercial and demonstration scales. Among the various technologies used for simultaneous waste reduction and biofuel production, thermochemical conversion which utilizes high heat to break down the organic matter component of waste materials into high value products has become increasingly attractive. For instance, due to biofuels technology advancement, current global electricity generation from biomass/waste conversion technologies (about 750

 
C. Ofori-Boateng, *Sustainability of Thermochemical Waste Conversion Technologies*,
https://doi.org/10.1007/978-3-031-64342-2_4

TWh) is projected to reach about 100 Exajoule (EJ) by 2050 due to their low emission properties and other advantages (Lee et al. 2022; Sikarwar et al. 2017).

Conventional thermochemical conversion technologies especially gasification have been in existence for close to 100 years where wood biomass was the primary source of feedstock to generate energy to power machines, cars, and thermal plants (Sikarwar et al. 2017). Since World War II, there has been tremendous development in the design of gasifiers and optimization studies of gasification and other thermochemical conversion processes for continuous improvement. For instance, the escalating progress of pyrolysis, gasification, combustion, hydrothermal liquefaction to mention only a few, has facilitated the possible integration of thermochemical waste conversion technologies with coal gasification facilities (like power plants) to improve sustainability (Huber et al. 2006). For instance, plasma gasification and pyrolysis discussed in Chapter 3 have been explored for decades but they are gaining new attention, interest, and investments as alternatives for waste management and biofuel production. This chapter discusses the recent advancement in thermochemical waste conversion and the way forward. Moreover, new, and innovative processes and techniques that are advantageous to traditional or conventional thermochemical conversion technologies are discussed. Conventional thermochemical waste conversion technologies like pyrolysis, incineration, combustion, etc., have been technically improved in so many ways including the integration of advanced process control systems and automation techniques. Recently, most waste-to-energy plants incorporating thermochemical conversion technologies use effective real-time monitoring tools for data analysis, maximizing product yields and reducing downtime whilst improving overall performance.

Nonetheless, as new developments in reactor design, process optimization, and sustainability analysis evolve, thermochemical waste conversion technologies are not devoid of challenges pertaining to thermodynamic, environmental, socio-economic, technological, and political sustainability. In view of this, thermochemical waste conversion technologies have not realized their full potential over the years. This chapter outlines the major challenges faced with technologically advanced thermochemical waste processes and the way forward.

## 4.2 Recent Development in Thermochemical Waste Conversion Technologies

The growing demand for alternative clean fuel and energy has propelled the need for continuous improvement of thermochemical conversion technologies which are effective ways of managing waste. Recent advancement in gasification technologies to improve product yields, utilize mixed waste materials, and reduce emissions are gradually drawing thermochemical conversion technologies closer to meeting the net zero goal of reducing $CO_2$ emissions by 2050.

Again, advancement in pyrolysis has recently focused on increasing energy efficiencies and reducing cost of production. Compared to previous years, thermochemical conversion technologies have seen massive increase in scale-up from laboratory research to pilot, demonstration, and commercial plants in most parts of the world (see Chapter 2) which includes commercial scale waste-to-energy facilities for the generation of green energy. The recent development of these technologies is pushing the economies forward whilst reducing the dependency on fossil fuels which would exhaust in the near future.

### 4.2.1 Machine Learning (ML) Applications

Research and development in thermochemical waste conversion technologies usually call for accurate experimental data to pinpoint optimal process conditions. Many conventional modeling techniques (e.g., kinetics and thermodynamic models, computational fluid dynamics, life cycle assessment, etc.) have evolved to facilitate fast data acquisition and interpretation during optimization studies to reduce time and cost (Gwak et al. 2020). However, these techniques are mostly difficult and unrealistic to apply to thermochemical conversion technologies because they are mostly based on assumptions which may impact the quality of model prediction leading to inconsistencies and difficulties in implementation of the results (Wong et al. 2013).

Machine learning has become one of the effective tools in analyzing a vast variety of system components and estimate characteristics of products. Machine learning is a subbranch of artificial intelligence (AI) which can act as a research tool for research question formulation, data collection, data selection, modeling, optimization, and data interpretation in a more efficient and reliable ways compared to the conventional modeling methods of analyzing research data (e.g., statistical analysis tools). It consists of a set of algorithms that are designed to model input data (Liu et al. 2021). Machine learning can be used to predict thermochemical conversion products' yields and process conditions. Machine learning algorithms for thermochemical waste conversion technologies according to Sun et al. (2023) include:

- ***Linear regression***, which is simple and easy to comprehend, has fast computation abilities and good performance on linearly separated data. However, the inability to manage non-linear relationships thus assumes linearity between features, coupled with its sensitivity with outliers puts them on the disadvantaged side.
- ***Neural networks***, which have the ability of studying complex patterns with good performance on big datasets. Nonetheless, they require considerable amount of data, long training time, many hyperparameters tuning, and they have high ability to overfit.
- ***Decision trees***, which are scalable to large datasets, easy to interpret and can manage both numerical and categorical data. Nevertheless, just like neural

network algorithms, decision trees also have high tendency to overfit, they are unstable and sensitive to noise.

- ***Random forest***, which can eliminate or reduce overfitting by merging multiple trees and helps in identifying critical process parameters of thermal treatment of waste. On the contrary, they are more complex and difficult to interpret with multiple trees unlike decision trees algorithms.
- ***Support vector machines***, which are suitable for high dimensional spaces and non-linear data and can be used with flexibility to choose suitable kernel functions. However, just like neural network machine learning algorithms, they have long training times for large datasets and become challenging when choosing kernel functions and parameters for non-linear problems.
- ***K-Nearest Neighbor***, a machine learning algorithm which is simple and easy to understand and not sensitive to outliers like linear regression algorithm. Yet, they have poor performance with high dimensional data, slow prediction speed, and requirement for large storage space.

Machine learning can help study the complex relationships and chemistry of waste materials, process conditions, and characteristics of the products formed during thermochemical waste conversion (Naqvi et al. 2023). Although there are many developed models that have shown good performance in literature, there are some waste materials and thermochemical conversion processes that machine learning methods can help reduce repetitive experiments in optimization studies and provide guidance for sustainable production in the real world (Sun et al. 2023).

Machine learning models can describe the behavior of waste thermochemical conversion processes based on data of simulation software such as ASPEN which can easily suggest conditions for optimum conversion processes. The activation energy of thermal treatment of waste can be predicted by machine learning models such as neural networks by considering waste compositions and process conditions simultaneously to ascertain the roles of process temperatures (Kartal and Özveren 2022).

Machine learning models are used to maximize the selectivity or yield of products and minimization of by-products. For instance, during gasification, tar concentration in syngas and total tar yield were predicted well (with coefficient of determination $R^2 = 0.988$) based on machine learning models (Serrano and Castelló 2020; Kardani et al. 2021). The roles of catalysts in pyrolysis of municipal sewage waste were investigated by Zaker et al. (2019) via simulation of the process using artificial neural networks algorithm using both catalytic and non-catalytic pyrolysis data. Their results confirmed the positive catalytic impact of sewage sludge char on pyrolysis with accurate prediction ability of the machine learning model ($R^2 > 0.99$) (Zaker et al. 2019).

### 4.2.2 Supercritical and Subcritical Water Thermochemical Waste Conversion

Thermochemical waste conversion technologies like gasification, pyrolysis, and combustion have been improved using supercritical and subcritical water treatments (discussed in Chapter 3). Recent development in supercritical water gasification or pyrolysis converts solid and liquid wastes (both hazardous and non-hazardous) into high energy gas with insignificant char formation (Modell 1985). Modell's research finding has triggered the development of several pilot, demonstration, and commercial plants around the world that use thermochemical conversion technologies coupled with supercritical or subcritical water treatment. Many authors have described super-, sub-, and near critical water properties in detail and some advantages of these technologies are listed below (Savage 1999; Kruse 2008):

- Because supercritical and subcritical waster gasification/pyrolysis operates at dense medium at high pressures (22.1 MPa), the reactor volume can be reduced to be more compact hence less heat loss.
- Because they are highly reactive and have catalytic activities, there is no need for catalysts (or if required, at low amounts) which makes the technology less costly in maintenance and operation.
- Supercritical and subcritical water gasification/combustion/pyrolysis allow the reaction to occur in a single phase hence there are no limitations to mass transfer as with heterogenous reactions where the reactants/products are in different phases.
- The high pressure supply with supercritical and subcritical water treatment is achieved with pump instead of gas via compressors, hence reducing the mechanical power requirement.
- With this technology, $CO_2$ can efficiently be removed and stored whilst the $H_2$ generated can be available at high pressure. Carbon capture technology can be applied at this point during the production of electricity and high purity hydrogen.
- Metals in the wastewater stream can be separated under supercritical conditions using appropriate separators.

### 4.2.3 Microwave-Assisted Thermochemical Waste Conversion

A microwave is a region between the infrared and radio waves (wavelengths of 0.01 to 1 m in the electromagnetic spectrum with frequencies ranging from 0.3 to 300 GHz (Motasemi and Afzal 2013). Microwave applications in industries for heating are usually between 915 and 2450 MHz (Motasemi and Afzal 2013). Microwave-assisted thermochemical conversion technology combines microwave radiation with conventional thermochemical conversion technologies like pyrolysis, gasification, combustion, etc., to convert electromagnetic energy into kinetic energy (Motasemi and Afzal

2013). During microwave heating, microwave energy is absorbed by the waste material and converted into heat within the materials which is then released outside thus facilitating fast and uniform continuous heating (Zhou et al. 2024).

Microwave-assisted thermochemical conversion of waste is one of the emerging innovative technologies highly efficient for mixed waste conversion into value-added products compared to conventional pyrolysis, gasification, combustion, etc. Microwave-assisted pyrolysis promoted secondary cracking of volatile fractions in pharmaceutical sludge pyrolysis gases and increased the content of methane, hydrocarbons, hydrogen, and CO via condensation, dehydrogenation, and aromatization (Zhou et al. 2024). The pyrolysis of mixed plastic waste assisted by microwave heating was found to generate high pyrolysis oil at 70 wt% yield with iron powder and coil ratio of 10:5 compared to conventional pyrolysis of plastic waste (Putra et al. 2024). The high performance of this process was attributed to the different shapes of metal (powder and coil) mixed with the plastic waste. The metal powder was able to absorb the microwave radiation whilst the metal coil concentrated the microwave at its surface and reflected it creating a synergistic effect that resulted in fast and efficient degradation of plastic components into pyrolysis oil (Putra et al. 2024). Variation in microwave power contributes to the nature of pyrolysis oil composition. For instance, at 300 W microwave power corresponding to 72 °C/min, plastic waste mix pyrolysis oil was found to contain aliphatic hydrocarbons (alkanes, alkenes, alkynes) suitable for gasoline substitute whilst 600 W power at high heating rate of 90 °C/min yielded pyrolysis oil with aromatic hydrocarbons which can be used as clean diesel substitute (Suriapparao et al. 2021). Other reports have concluded that bio-oils generated by microwave-assisted thermochemical conversion contain phenols, light hydrocarbons without polycyclic aromatic hydrocarbons which is contrary to that of conventional methods which generate products with hydrophilic and corrosive oxygen (Halim and Swithenbank 2016; Robinson et al. 2007).

Microwave-assisted pyrolysis of different types of straws (namely rice straw, corn straw, and wheat straw) at 400 °C and microwave power of 33.3 kW per kg waste for 25 min resulted in an optimal syngas and hydrogen yields of 58.7% and 25.4% respectively (Yue et al. 2024). However, with catalytic microwave-assisted pyrolysis (NiO/C composites) at same conditions, the total gas yield improved significantly at about 1.9 folds higher than non-catalytic microwave-assisted pyrolysis (Yue et al. 2024).

Microwave gasification is also found to generate high hydrogen yield and enhanced selectivity to syngas over tar (Abedin et al. 2023). Mixed plastics and corn stover were gasified under microwave conditions at 700 °C yielding about 30.5 mmol $H_2$ per gram waste which was about 30 times higher than conventional heating via gasification (Abedin et al. 2023). In a similar research, air gasification of rice husks in a fluidized bed reactor assisted with microwave radiation at 900 °C resulted in high syngas yield of 78.2 wt%, cold gas efficiency of 81.8%, and higher heating value of 6.3 $MJ/Nm^3$ with significantly low tar formation (2.1 wt%) compared to conventional catalytic gasification at 950 °C (Fu et al. 2023). Microwave-assisted catalytic gasification can utilize all kinds of catalysts and cheap absorbers and still generate biochar with high quality (Fang et al. 2021). Absorbers are dielectric materials which are usually blended with the biomass[1] to increase heat transfer within the materials for efficient organic matter degradation.

It is reported that biochar generated from microwave-assisted pyrolysis possessed high surface area, high pore volume, high uniformity, and cleanness compared to biochar from conventional pyrolysis (Zaker et al. 2019; Huang et al. 2016).

### 4.2.4 Ultrasonic-Assisted Thermochemical Waste Conversion

Ultrasound consists of mechanical waves with frequencies of 20 kHz–10 MHz which need an elastic medium to grow. Ultrasound region is characterized by power ultrasound (for industrial processes like thermochemical conversion at frequencies of 20–100 kHz) and signal or diagnostic ultrasound (for medical diagnostics at frequencies of 100 kHz–10 MHz) (Tiwari 2015; Yang et al. 2020). The technology of ultrasound based on the principle of acoustic cavitation (which progresses from formation, growth, and collapse of bubbles in liquid media) in enhancing the efficiencies of thermochemical waste conversion has evolved as one of the effective ways to generate value-added bioproducts from waste due to their excellent merits compared to conventional pyrolysis, gasification, combustion, etc. (Vázquez-Rodríguez et al. 2020). Ultrasound bubble release intense flow of matter and energy when they explode thus the effect changes the surface tension of fluid which intends directs the energy and matter flow toward the waste material' surface generating strong mechanical effect up to about $10^3$ MPa (Fang et al. 2023). Ultrasound treatment of waste is reported to manifest mechanical impacts at low frequencies of 20–60 kHz and sono-chemical effects at higher frequencies (100–1000 kHz) (Yang et al. 2020).

Zeolite-based MO/HZSM-5 catalyst modified via ultrasonic treatment to enhance the efficiency of pyrolysis and reduce the coking issues was found to increase aromatic carbon production by about 54% compared to conventional catalytic pyrolysis of waste biomass (Yu et al. 2023).

Long residence time facilitates the formation of high levels of phenols and alcohols whilst short residence time increases syngas formation with char yields decreasing during ultrasonic-assisted pyrolysis of waste. In an experiment where bagasse and sludge were co-pyrolyzed using ultrasound frequency of 45 kHz at 200 W for 2 h, char formation was low and gas formation was high compared to 1 h of ultrasound treatment at same conditions which led to phenols and alcohols formation (Fang et al. 2023). Table 4.1 summarizes the pros and cons of microwave heating and conventional thermal treatment of waste materials.

[1] Some waste does not have high absorbing capabilities, they are usually blended with microwave absorber that are made from special types of rubber or foam absorb microwave for efficient heat transfer (Ethaib et al. 2020).

**Table 4.1** Differences between microwave, ultrasonic, and conventional thermal treatment of waste materials for fuel and energy production (Frediani and Frediani 2022)

| Technology | Advantages | Disadvantages |
|---|---|---|
| Conventional Heating | • Uses diverse fuel sources<br>• May or may not require catalysts or additives | • Heats from the surface of the material<br>• Long residence time<br>• Low heating efficiency<br>• Combustion gases may contaminate product gases<br>• Difficult heating of complex polymers in waste |
| Microwave | • High heating efficiency<br>• Short residence time<br>• Product gas does not contain combustion gases<br>• Easy heating of complex polymers | • Requires microwave absorber (e.g., activated carbon, graphite, silicon carbide) especially for waste material with low dielectric constant<br>• Electricity is needed |
| Ultrasonic-assisted heating | • Few instrumental requirements<br>• Low economic and environmental impacts<br>• Low solvent volume | • High energy requirement<br>• High capital cost |

### 4.2.5 *Waste-To-Energy (WtE) Hybrid Systems with Renewable Energy Systems*

As discussed in Chapter 3, one major development and improvement on sustainability of thermochemical conversion technologies is hybrid or integration systems of renewable energy like solar, wind, etc., with thermochemical conversion technologies like pyrolysis, gasification, etc. Thermochemical conversion of waste materials to bioenergy (bioelectricity and heat) can occur in an integrated system which are renewable energy technologies (such as solar thermal, photovoltaics, fuel cells, anaerobic digestion, etc.). These integrated systems are called thermochemical hybrid waste-to-energy systems (THS), and they have evolved as one of the sustainable strategies of reducing waste and $CO_2$ emissions into the atmosphere.

THS can increase energy storage capacities, improve energy efficiency, and power quality hence reducing energy cost compared to single-source renewable energy systems like photovoltaic (Wang et al. 2015). Also, according to Ashok (2007), THS can reduce environmental impacts, reduce electrical lines and transformer accidents, and facilitate efficient transmission and distribution of power hence rendering them dependable and safe.

### 4.2.6 *Products and By-Product Upgrade and Utilization*

Though the products from thermochemical conversion technologies can be substitutes of gasoline, hydrogen, diesel, methanol, etc., they need to be upgraded to improve the quality and promote the sustainability of the technologies. There are many methods of products and by-product upgrades of bio-oil, syngas, biochar, ash, etc., which include the following:

- ***Fischer–Tropsch synthesis (FTS)*** is a catalytic means of converting syngas into liquid hydrocarbons and oxygenates, and it is used to upgrade low grade product gases from thermochemical conversion of waste into high-quality fuels and chemicals. Chapter 2 and Chapter 3 of this book discussed the global statistics and application of products generated through FTS with thermochemical conversion technologies, respectively.
- ***Functionalization*** is a novel way of increasing the range of applications of bio-oil generated from thermochemical conversion technologies by chemical modification, where functional groups from other products are added to crude bio-oil to improve their properties and compatibility for specific chemical reactions and diverse application in industries. Hydrogenation is a method of functionalization where hydrogen is added to syngas or bio-oil to improve its stability, energy content, etc. Other functionalization methods for bio-oil and syngas upgrade include esterification, oxidation, alkylation, etc. Deoxygenation can be employed to remove oxygen atoms from bio-oil and render them more suitable for fuels and chemicals.
- ***Catalytic upgrading*** of crude bio-oil makes them useful in diesel and gasoline engines. This upgrade method utilizes catalyst to facilitate chemical reactions to modify the composition of syngas and bio-oil to achieve high product quality and yield, high selectivity, and process efficiency.
- ***Fractionation*** of bio-oil via fractional distillation where the oil is heated to vaporize the various components by boiling point differences before condensing into liquid. Syngas fractionation involves the separation of CO and $H_2$ by cryogenic distillation where the gases are cooled to low temperatures for efficient separation. The resulting products include high-quality biofuels and biochemicals which have diverse applications in transportation, thermal plants, and chemical industries.
- Low-grade bio-oil can be refined via ***cracking*** to eliminate impurities and improve the quality.

The residual ash from waste incineration and combustion can be used for many purposes after the phosphorus has been effectively removed. Random forest machine learning modeling was efficiently applied to optimum phosphorus removal from incineration ash with close relations to the origins and processing conditions (Joseph et al. 2022; Fang et al. 2021).

### 4.2.7 Emissions Reduction Techniques and Carbon Dioxide Capture

Air pollutants such as dioxins from incinerators that are not controlled are detrimental to humans and the environment at large. Though incineration is found to be one of the commonly used thermochemical conversion technologies for waste management, emissions are one of the major drawbacks especially in countries where $NO_x$ emission limits have been established for incineration systems. Over the years, combustion and controlled incinerations have integrated waste-to-energy systems to capture the gases and heat for electricity generation. More recently, some methods for efficient particulate matter removal have evolved where modern municipal solid waste (MSW) incinerators have fabric filters or dry electrostatic precipitators (ESP). For the removal of dioxins and mercury, dry powdered activated carbon injection systems, spray dryer absorbers, and dry-lime injection systems have significantly reduced acid-based emissions and heavy metals into the environment.

*Carbon dioxide ($CO_2$) capture* and storage facilities aim to prevent $CO_2$ generated from industrial activities and power plants from being released into the atmosphere. $CO_2$ capture is a negative emissions technology which removes and stores $CO_2$ from the atmosphere hence reducing greenhouse gas (GHG) emissions. An example of a negative emissions technology that combines biomass conversion for energy (waste-to-energy) and $CO_2$ capture is bioenergy with carbon dioxide capture and storage (BECCS) which has been extensively studied (Cheng et al. 2021). The feasibility of thermochemical waste (e.g., MSW, agricultural residues, etc.) conversion technologies (like pyrolysis, gasification, combustion, hydrothermal treatment, etc.) being integrated with BECCS technologies have been investigated (Cheng et al. 2021; Shahbaz et al. 2021). Many countries like Japan, China, United Kingdom, United States, etc., have adopted bioenergy with carbon dioxide capture and storage into their combustion and incineration facilities (Shahbaz et al. 2021). For instance, a demonstration plant in Japan using coal and biomass for power generation captures about 180,000 $CO_2$ annually (Saito et al. 2014). In the United States, carbon sequestration and economic burdens of net emission technologies were assessed for wood waste combustion with carbon capture and were found to contribute about 98% of the negative carbon sequestration effectively (Cheng et al. 2021). Hydrothermal treatment and slow pyrolysis of waste was found to have small contributions to carbon sequestration because of small market size of char and high levelized cost of negative $CO_2$ emissions for hydrothermal treatment technologies (Cheng et al. 2021). Though waste-to-power plants integrated with $CO_2$ capture and storage are found to be environmentally and economically beneficial, there remains room for improvement as there are drawbacks.

### *4.2.8 Thermochemical Conversion of Wastes Involving Co-Production*

A thermochemical conversion process whereby two or more feedstocks are used as raw material for energy, fuel, and chemicals production in one reactor is termed co-production or co-processing. Different kinds of waste have been mixed and subjected to thermal decomposition successfully where high yield bio-oil, biochar, and syngas have been generated (Sewu et al. 2019; Diao et al. 2021; Johannes et al. 2013). The successful co-pyrolysis, co-combustion, co-gasification, and co-incineration (the commonly developed technologies with co-production) required synergistic effects through series of radical interactions[2] because of contact of different waste materials. The physicochemical composition of the mixed waste material as well as process conditions and catalysts during co-pyrolysis, co-combustion, and co-gasification would impact process efficiency hence products yield and quality.

Co-pyrolysis is found to enhance the yield and quality of biochar without modification to the pyrolysis process except with the feed preparation unit (Yao et al. 2021). Rodriguez et al. (2021) concluded that co-pyrolysis of agricultural residues mixed with industrial waste improved the electrical conductivity, stability, water-holding capacity, and neutralizing power of biochar generated hence highly suitable for fertilizer and soil remediation applications.

## 4.3 Drawbacks of Thermochemical Waste Conversion Technologies

### *4.3.1 Technological Complexity and Waste Composition Challenges*

Despite recent advancement in waste management using thermochemical conversion technologies, there are considerable number of challenges that affect the sustainability of waste conversion including the heterogeneity of mixed waste composition, high quantity of reaction solvents and media required, dependence on high pressure and temperature, reactor design problems and calorific value loss resulting from inorganic contamination to mention but a few (Pandey et al. 2015). For instance, for waste combustion, the variation of composition and physicochemical properties of the waste give rise to difficulties in adjusting the combustion operation parameters (Wang and Wu 2023). There is the limitation of high oxygen and moisture contents

---

[2] Series of radical reactions occurring during co-production include initiation, formation of secondary radicals (depolymerization, formation of monomers, favorable and unfavorable hydrogen transfer reactions, intermolecular hydrogen transfer, isomerization) as well as termination by disproportionation or recombination of radicals. These reactions have been detailed by Önal et al. (2014).

in the waste materials during combustion which may lead to low calorific value of the waste hence the need for more waste materials resulting in high operation cost (Wang and Wu 2023). According to Wang and Wu (2023), the main limitations of torrefaction technology include:

- Self-heating and self-ignition of biochar generated due to the slow rate of heat removal from the reactor.
- Compared to coal torrefaction, waste torrefaction has high investment and operating cost
- Ash-related issues in boilers and furnaces which may cause slagging, agglomeration, and corrosion of equipment hence low process efficiency.

The drawbacks of MWS incineration are normally associated with incomplete burning in the combustion chamber as well as low energy conversion. These challenges can be overcome by innovations and process continuous improvement through co-combustion with other waste materials (Sun et al. 2022). Chapter 3 of this book discussed the feasibility of producing bioproducts from mixed waste using co-pyrolysis, etc. Though this can effectively be done, there is more room for improvement to study the unclear reaction mechanisms in co-pyrolysis, co-combustion, etc., to eliminate the trial-and-error method and reduce cost making thermochemical conversion technologies sustainable in the long term (Ni et al. 2022). As a result, machine learning is used to model the process to show good agreement between the experimental data and predicted ones (Chen et al. 2018).

Though waste-to-energy via thermochemical conversion processes integrated with other renewable energy systems have economic and environmental benefits, there are some challenges that remain such as the following:

- Because hybrid system for thermochemical conversion of waste utilizes diverse waste in significant amount, the total amount of waste generated in the world may not be enough to meet the energy demand in some parts of the world.
- Lack of social awareness of hybrid waste-to-energy systems may hamper the growth of these facilities. There may be some potential investors who can support these hybrid system, but they do not believe in green energy. Sensitization and promotion of biofuels must be done across all sectors of the world to promote the use of biofuel and displace fossil-based fuels and chemicals.
- Despite the proposal and promotion of hybrid waste-to-energy systems, the designs are mostly conceptual with few studies having experimental validation.

The major drawback with catalytic pyrolysis of waste over zeolite-based catalyst (which is commonly used for catalytic thermochemical waste conversion) is the significant generation of by-products and fast coking of catalyst (Yu et al. 2023). Conventional pyrolysis also has the drawback of generating viscous and impure pyrolysis oil which may not be directly utilized in automobile engines without modifications hence the need to upgrade the oil. Also, there is a problem with catalyst deactivation as a result of coke deposition (Wang and Wu 2023).

Similarly, gasification of waste into syngas and biochar poses the challenge of catalyst deactivation caused by sintering and coke deposition which reduce reaction rates and increase cost of operation (Wang and Wu 2023). As mentioned earlier, tar and soot formation during gasification is still a hurdle for the thermochemical conversion technology industry especially with plastic waste gasification due to repolymerization and interactions of tar with other contaminants, blockage of filters, and condensation on cooler surfaces causing operation issues (Palma 2013). Above all, large-scale operation and maintenance of gasification systems are costly due to high cost of capital investment.

Microwave-assisted thermochemical conversion technologies are efficient but with some drawbacks. The energy conversion efficiency is low in the range of 20–60% and the interference between electromagnetic irradiation and thermocouple sensors also hinders accurate data acquisition in the reactor (Yin 2012). Again, the scale-up of microwave-assisted thermochemical conversion is difficult due to the limitation of microwave penetration deep into the material for uniform heat transfer (Sun et al. 2016).

For hydrothermal liquefaction of waste, complexity of technology, operation conditions (high pressure), and additives as well as cost may limit commercial scale operation. Also, the composition of liquid fuel produced (also with low calorific value) during hydrothermal liquefaction is more complicated compared to gasoline and diesel hence they may need further upgrade before utilization in automobile engines (Wang and Wu 2023).

### 4.3.2 High Capital and Operational Costs

Thermochemical conversion technologies are energy-intensive due to the high-temperature requirements to effectively decompose the organic matter in the waste. Hydrothermal treatment (see Chapter 3) can be used to offset the energy-intensive drying and convert the waste directly into value-added bioproducts. Hot compressed water from waste which contains a lot of water (like sewage sludge, food waste, etc.) can function as the reaction media or catalyst in hydrothermal treatment (like hydrothermal gasification, hydrothermal pyrolysis, hydrothermal liquefaction, etc.) to break down the large hydrocarbon molecules into smaller ones (Zhu et al. 2023). Thermochemical conversion technologies are expensive when it comes to equipment and facilities which make the capital cost huge and maintenance and parts replacement costly as well.

Another drawback of thermochemical conversion processes is the lack of identical framework or data on efficient recovery of by-products on laboratory, pilot, and commercial scale which makes selection of a suitable thermochemical conversion technology difficult. Also, waste materials with complex organic and inorganic components (like MSW) as well as the optimization of process conditions with complex reaction mechanisms make thermochemical conversion processes difficult for continuous improvement.

### 4.3.3 Sustainability and Sustainable Development Challenges

The adoption of sustainability standards within thermochemical waste conversion facilities has become difficult due to disparities in biofuel standards from one country to another. Sustainability indicators and criteria must be developed to specifically address the challenges of small- and large-scale producers of bioenergy and biofuels.

*High Certification Cost* for small-scale producers has become another challenge in adopting sustainability and sustainable development in these facilities. Government assistance and private partnership among small-scale biofuel producers would help overcome certification barriers in terms of finance and technology by providing good incentives.

*Capacity Building* is another challenge in the thermochemical conversion industries for sustainability adoption. The lack of sufficient knowledge on green technologies is the problem. Training programs for biofuel producers can help lots of producers to achieve sustainability.

*Lack of External Support* from financial organizations, investors, importers, retailers, lenders, downstream processers could affect the sustainable production and consumption of biofuels generated via thermochemical conversion technologies.

### 4.3.4 Policies and Regulatory Challenges

Thermochemical conversion technologies may have restrictions from government which may require permits, certification, etc., hence increasing the cost of production and delaying production. Policies and regulatory challenges of thermochemical conversion technologies for biofuel production are applied to emissions, permit requirements, subsidies and incentives, waste collection and utilization requirements, land use and zoning regulations.

Government may impose strict regulations and laws on emissions from thermal conversion facilities to protect the ecosystem. However, biofuel producers may not be able to meet these strict standards which can limit profitability of thermochemical conversion facilities. The permits required by thermochemical conversion facilities for operation undergo stringent environmental reviews and compliance which deter investors. Biofuel producers and investors who undergo such reviews go through costly processes which may affect the overall cost of product hence sustainability of the process.

Again, though subsidies and incentives are positive biofuel development criteria, they pose many challenges to investors and producers. Provision of subsidies and incentives for biofuels can disturb the fuel market by inflating demand for biofuels artificially which can create instability in the market and eventually destroy industries who are not receiving incentives. Also, subsidies and incentives are usually accompanied by strict conditions such as emissions limits into the environment. Thermochemical conversion technologies like incineration, combustion, and those which release considerable amount of emissions may be exacerbated by these rules governing subsidies and incentives. Government subsidies can also result in international trade disputes which can affect importers and exporters of biofuels. Currently, there are government subsidies which support only a certain type of biofuels especially biodiesel from first and second generation feedstock (biochemical conversion technologies). This may discourage research and development in other areas such as thermochemical conversion technologies. Thus, it is imperative to carefully understand the underlining conditions of incentives and subsidies to avoid unintended outcomes.

Also, some thermochemical conversion technologies may be subject to land use and zoning regulations which usually restrict the location permitted to install facilities.

The commercialization of some thermochemical conversion technologies like plasma gasification remains difficult for investors because of the absence of complete policy system as well as high level of operational skills required for the process to run efficiently (Ayub et al. 2023).

### *4.3.5 Emissions and Environmental Burdens*

Despite the recent development in carbon capture and emissions mitigation with thermochemical conversion technologies, there is still room for improvement, especially for combustion and incineration technologies. The generation of air pollutants, greenhouse gases, and particulate matter affects the environmental sustainability of thermochemical conversion technologies. The collection, sorting, and transportation of waste materials for thermochemical conversion can have environmental burdens which can affect the sustainability of production facilities.

## 4.4 Breakthroughs of Thermochemical Waste Conversion Technologies

Bioenergy and biofuels from thermochemical conversion technologies are expected to increase in generation and consumption capacities due to breakthroughs in policy incentives and government subsidies. There is currently an increasing shift in biofuels policies toward sustainable energy production at low carbon footprint. In the United Kingdom for instance, the Climate Change Act 2008 is set in place to drastically reduce 80% of carbon emission in the United Kingdom by 2050 (Climate Change Act 2008). As a result, there is a Renewable Obligation Certificate which the UK government has provided as subsidies to support the low carbon energy generation, and this has moved the UK's biofuel market forward where about 15 waste-to-energy plants are in operational. In the United States for instance, the renewable fuel Standard

(RFS) is expected to sustain biofuel production and demand though there are certain constraints associated with the implementation of this policy (OECD/FAO 2020).

The increasing awareness of climate change and sustainability coupled with the near depletion of fossil fuel reserves have significantly increased the drive in producing more cleaner fuels and chemicals. As discussed in Chapter 2 of this book, biochar, bio-oil, and syngas production market has significantly expanded over the years.

Also, collaboration and partnerships between research institutions and universities, industries and other government agencies have enhanced knowledge-sharing and joint research endeavors. This has led to the rapid growth and development of thermochemical conversion technologies.

Feedstock availability for waste-to-energy technologies may never be an issue as the global generation of waste increases day by day (see Chapter 1) unlike other technologies (biochemical conversions) which require first generation feedstock (e.g., corn, canola oil) that compete with global food consumption.

## 4.5 Conclusion

Though a larger percentage of global biofuel production and consumption are sourced from thermochemical conversion technologies, there are still limitations that may have to be resolved in order to achieve sustainability. Government policy and regulations limitations, high capital and operational costs, technological complexity and emissions are the few drawbacks of thermochemical waste conversion technologies. Nonetheless, there are major breakthroughs that have been realized over the years for pyrolysis and gasification technologies for instance. This chapter elaborated on the main sustainability challenges and solutions, current technological development, and prospects of thermochemical conversion technologies for biofuels, bioenergy, and biochemicals production.

## References

Abedin A, Bai X, Smith M, Muley P (2023) Microwave-assisted co-gasification of mixed plastics and corn stover: A synergistic approach to produce clean hydrogen. Energy Convers Manag 280:116774

Ashok S (2007) Optimised model for community-based hybrid energy system. Renew Energy 32:1155–1164

Ayub Y, Zhou J, Ren J, Wang Y, Shen W, He C, Dong L, Toniolo S (2023) Plasma gasification based monetization of poultry litter: System optimization and comprehensive 5E (Energy, Exergy, Emergy, Economic, and Environmental) analysis. Energy Convers Manag 282:116878

Chen J, Xie C, Liu J, He Y, Xie W, Zhang X et al (2018) Co-combustion of sewage sludge and coffee grounds under increased O(2)/CO(2) atmospheres: Thermodynamic characteristics, kinetics and artificial neural network modeling. Bioresour Technol 250:230–238

Cheng F, Small AA, Colosi LM (2021) The levelized cost of negative CO2 emissions from thermochemical conversion of biomass coupled with carbon capture and storage. Energy Convers Manag 237:114115

Climate Change Act (2008) Chapter 27 of 2008. United Kingdom. 2016. http://www.legislation.gov.uk/ukpga/2008/27/pdfs/ukpga_20. Accessed 15 March 2024

Diao R, Sun M, Huang Y, Zhu X (2021) Synergistic effect of washing pretreatment and co-pyrolysis on physicochemical property evolution of biochar derived from bio-oil distillation residue and walnut shell. J Anal Appl Pyrolysis 155:105034

Ethaib S, Omar R, Kamal SMM, Awang Biak DR, Zubaidi SL (2020) Microwave-assisted pyrolysis of biomass waste: a mini review. Processes 8(9):1190

Fang L, Wang Q, Li J-S, Poon CS, Cheeseman C, Donatello S, Tsang DC (2021) Feasibility of wet-extraction of phosphorus from incinerated sewage sludge ash (ISSA) for phosphate fertilizer production: a critical review. Cri Rev Env Sci Tec 51(9):939–971

Fang S, Li C, Yan S, Huang Z, Lin Y, Zou H, Zhang L (2023) Ultrasonic pretreatment effect on the co-pyrolysis characteristics and products of bagasse and municipal sludge. J Energy Inst 111:101240

Frediani P, Frediani M (2022) Mixed or Contaminated Waste Plastic Recycling through Microwave—Assisted Pyrolysis. IntechOpen. https://doi.org/10.5772/intechopen.100179

Fu W, Zhang Y, Cui L, Liu H, Maqsood T (2023) Experimental microwave-assisted air gasification of biomass in fluidized bed reactor. Bioresour Technol 369:128378

Gwak YR, Yun JH, Keel SI, Lee SH (2020) Numerical study of oxy-fuel combustion behaviors in a 2MWe CFB boiler. Korean J Chem Eng 37(11):1878–1887

Halim SA, Swithenbank J (2016) Characterisation of Malaysian wood pellets and rubberwood using slow pyrolysis and microwave technology. J Anal Appl Pyrolysis 122:64–75

Huang Y-F, Chiueh P-T, Lo S-L (2016) A review on microwave pyrolysis of lignocellulosic biomass. Sustain Environ Res 26(3):103–109

Huber GW, Iborra S, Corma A (2006) Synthesis of transportation fuels from biomass: chemistry, catalysts, and engineering. Chem Rev 106:4044–4098

Johannes I, Tiikma L, Luik H (2013) Synergy in co-pyrolysis of oil shale and pine sawdust in autoclaves. J Anal Appl Pyrolysis 104:341–352

Joseph C-A, Khiari L, Gallichand J, Beecher N (2022) Assessing and predicting phosphorus phytoavailability from sludge incineration ashes. Chemosphere 288:132498

Kardani N, Zhou A, Nazem M, Lin X (2021) Modelling of municipal solid waste gasification using an optimised ensemble soft computing model. Fuel 289:119903

Kartal F, Özveren U (2022) Prediction of activation energy for combustion and pyrolysis by means of machine learning. Therm Sci Eng Prog 33:101346

Kruse A (2008) Supercritical water gasification. Biofuel Bioprod Biorefin 2(5):415–437

Lee D, Nam H, Seo MW, Lee SH, Tokmurzin D, Wang S, Park Y-K (2022) Recent progress in the catalytic thermochemical conversion process of biomass for biofuels. Chem Eng J 447:137501

Liu Y, Esan OC, Pan Z, An L (2021) Machine learning for advanced energy materials. Energy AI 3:100049

Modell M (1985, September) Processing methods for the oxidation of organics in supercritical water. US Patent 453190

Motasemi F, Afzal MT (2013) A review on the microwave-assisted pyrolysis technique. Renew Sustain Energy Rev 28:317–330

Naqvi SR, Ullah Z, Taqvi SAA, Khan MNA, Farooq W, Mehran MT, Juchelková D, Štěpanec L (2023) Applications of machine learning in thermochemical conversion of biomass—a review. Fuel 332:126055

Ni ZS, Bi HB, Jiang CL, Sun H, Zhou WL, Tian JJ, Lin QZ (2022) Investigation of co-combustion of sewage sludge and coffee industry residue by TG-FTIR and machine learning methods. Fuel 309:122082

OECD/FAO (2020) OECD-FAO agricultural outlook 2020–2029. OECD Publishing, Paris/FAO, Rome.https://doi.org/10.1787/1112c23b-en.Accessed26Feb

Önal E, Uzun BB, Pütün AE (2014) Bio-oil production via co-pyrolysis of almond shell as biomass and high density polyethylene. Energy Convers Manag 78:704–710

Palma CF (2013) Modelling of tar formation and evolution for biomass gasification: a review. Appl Energy 111:129–141

Pandey A, Negi S, Binod P, Larroche C (eds) (2015) Pretreatment of biomass: processes and technologies. Elsevier B.V., Waltham, USA

Putra PHM, Rozali S, Patah MFA, Ghazali NNN, Ahmad R, Idris A (2024) Effect of metal powder and coil on microwave pyrolysis of mixed plastic. Chem Eng J 487:150530

Robinson JP, Kingman SW, Snape CE, Shang H (2007) Pyrolysis of biodegradable wastes using microwaves. Proc Inst Civ Eng-Waste Resour Manag, Thomas Telford Ltd, pp 97–103

Rodriguez JA, Lustosa Filho JF, Melo LCA, de Assis IR, de Oliveira TS (2021) Co-pyrolysis of agricultural and industrial wastes changes the composition and stability of biochars and can improve their agricultural and environmental benefits. J Anal Appl Pyrolysis 155:105036

Saito S, Udatsu M, Kitamura H, Murai S, Kato Y, Maezawa Y, Watando H (2014) Development and evaluation of a new amine solvent at the Mikawa $CO_2$ capture pilot plant. Energy Procedia 51:176–183

Savage PE (1999) Organic chemical reactions in supercritical water. Chem Rev 99(2):603–622

Serrano D, Castelló D (2020) Tar prediction in bubbling fluidized bed gasification through artificial neural networks. Chem Eng J 402:126229

Sewu DD, Lee DS, Tran HN, Woo SH (2019) Effect of bentonite-mineral co-pyrolysis with macroalgae on physicochemical property and dye uptake capacity of bentonite/biochar composite. J Taiwan Inst Chem Eng 104:106–113

Shahbaz M, AlNouss A, Ghiat I, Mckay G, Mackey H, Elkhalifa S, Al-Ansari T (2021) A comprehensive review of biomass based thermochemical conversion technologies integrated with $CO_2$ capture and utilisation within BECCS networks. Resour Conserv Recycl 173:105734

Sikarwar VS, Zhao M, Fennell PS, Shah N, Anthony EJ (2017) Progress in biofuel production from gasification. Prog Energy Combust Sci 61:189–248

Sun H, Bi H, Jiang C, Ni Z, Tian J, Zhou W, Qiu Z, Lin Q (2022) Experimental study of the co-pyrolysis of sewage sludge and wet waste via TG-FTIR-GC and artificial neural network model: Synergistic effect, pyrolysis kinetics and gas products. Renew Energ 184:1–14

Sun J, Wang W, Yue Q (2016) Review on microwave-matter interaction fundamentals and efficient microwave-associated heating strategies. Materials 9(4):231

Sun L, Li M, By L, Li R-H (2023) Machine learning for municipal sludge recycling by thermochemical conversion towards sustainability. Bioresour Technol 394(9):130254

Suriapparao DV, Nagababu G, Yerrayya A, Sridevi V (2021) Optimization of microwave power and graphite susceptor quantity for waste polypropylene microwave pyrolysis. Process Saf Environ Prot 149:234–243

Tiwari BK (2015) Ultrasound: a clean, green extraction technology. TrAC Trends Anal Chem 71:100–109

Vázquez-Rodríguez B, Gutiérrez-Uribe JA, Antunes-Ricardo M, Santos-Zea L, Cruz-Suárez LE (2020) Ultrasound-assisted extraction of phlorotannins and polysaccharides from *Silvetia compressa* (Phaeophyceae). J Appl Phycol 32:1441–1453

Wang X, Palazoglu A, El-Farra NH (2015) Operational optimization and demand response of hybrid renewable energy systems. Appl Energy 143:324–335

Wang Y, Wu JJ (2023) Thermochemical conversion of biomass: potential future prospects. Renew Sustain Energy Rev 187:113754

Wong KI, Wong PK, Cheung CS, Vong CM (2013) Modelling of diesel engine performance using advanced machine learning methods under scarce and exponential data set. Appl Soft Comput 13(11):4428–4441

Yang X, Cui C, Zheng A, Zhao Z, Wang C et al (2020) Ultrasonic and microwave assisted organosolv pretreatment of pine wood for producing pyrolytic sugars and phenols. Ind Crops Prod 157:112921

Yao Z, Kang K, Cong H, Jia J, Huo L, Deng Y, Xie T, Zhao L (2021) Demonstration and multi-perspective analysis of industrial-scale co-pyrolysis of biomass, waste agricultural film, and bituminous coal. J Clean Prod 290:125819

Yin C (2012) Microwave-assisted pyrolysis of biomass for liquid biofuels production. Bioresour Technol 120:273–284

Yu J, Luo B, Wang S, Wu K, Tao Y, Chu S, Zhang H (2023) Enhanced aromatic yield from catalytic pyrolysis of pine wood via ultrasonic assisted Mo modified HZSM-5. Biomass Bioenergy 172:106769

Yue M, Cheng J, Jiang Q, Xu G, Wang J, Fu Y, Ye F (2024) Enhanced hydrogen production from straws using microwave-assisted pyrolysis with NiO/C based catalyst/absorbent. Int J Hydrog Energy 59:535–550

Zaker A, Chen Z, Wang X, Zhang Q (2019) Microwave-assisted pyrolysis of sewage sludge: a review. Fuel Process Technol 187:84–104

Zhou Y, Lin F, Ling Z, Zhan M, Zhang G, Yuan D (2024) Comparative study by microwave pyrolysis and conventional pyrolysis of pharmaceutical sludge: resourceful disposal and antibiotic adsorption. J Hazard Mater 468:133867

Zhu X, Liu B, Sun L, Li R, Deng H, Zhu X, Tsang DCW (2023) Machine learning-assisted exploration for carbon neutrality potential of municipal sludge recycling via hydrothermal carbonization. Bioresour Technol 369:128454

# Part III
# Sustainability Assessment of Thermochemical Biomass Conversion Technologies

# Chapter 5
# Exergy and Life Cycle Analyses of Thermochemical Waste Conversion Technologies

**Abstract** Thermochemical conversion technologies produce eco-friendly products like biochar, bio-oil, and syngas which have a wide variety of applications in the transportation and chemical industries. Nonetheless, about 95% of the energy used to produce these bioproducts is sourced from fossil fuels which dissipate lots of energy and other emissions into the environment and eventually reduce the exergetic efficiencies of thermochemical conversion technologies. The adoption and implementation of sustainability criteria and principles in thermochemical conversion technologies as well as continuous process and equipment improvement would help reduce exergy destruction and environmental burdens. This chapter discusses the thermo-environmental sustainability of conventional and newly developed thermochemical conversion technologies like combustion, gasification, plasma pyrolysis, microwave-assisted pyrolysis, etc. via the exergy and life cycle assessment tools. Exergetic life cycle analysis (ExLCA) would help pinpoint locations of quality energy inefficiencies and evaluate the impacts of products and processes on the environment.

**Keywords** Life Cycle Assessment · Exergy Analysis · Sustainability · Exergetic life cycle analysis · Waste-to-energy · Thermochemical conversion · Waste management

## 5.1 Introduction

In the energy sector of the world's economy, the trending topics have not ceased to be discussed pertaining to carbon neutrality and energy insecurity due to the near exhaustion of fossil fuels coupled with the detrimental effects on the environment. As a result, bioenergy and biofuels which form only about 10–15% of global primary energy has gained much interest, and there are numerous pilot plants, demonstration, and commercial scale waste-to-energy plants being installed all over the world to help meet the 2050 carbon dioxide emission reduction target (IEA 2023). Again, since the cost of energy in most manufacturing companies including thermochemical waste conversion facilities is about 20–80% of the variable cost of the plant, it is

 
C. Ofori-Boateng, *Sustainability of Thermochemical Waste Conversion Technologies*,
https://doi.org/10.1007/978-3-031-64342-2_5

vital to minimize energy utilization and improve equipment performance to make these production facilities sustainable (Ayres et al. 2007; Rosen and Dincer 2003). According to the UN Environment Program (2019), about 36,000 million tons of $CO_2$ (concentration of about 407 ppm) were emitted into the atmosphere with about 2.8% annual increase due to fossil fuel use and anthropogenic activities.

The first law of thermodynamics depicts that energy is conserved for all processes, thus there is no degradation. According to the second law of thermodynamics, in all energy conversion processes there is a certain amount of useful energy that is dissipated into the environment and hence destroyed in the process. Thus, the operation of thermochemical conversion technologies which involves the transformation of biomass and waste materials into bioenergy (e.g., heat, electricity), biofuels, and biochemicals is not devoid of entropy[1] generation and thermodynamic inefficiencies. Irreversibilities and entropy generation during thermochemical conversion processes tend to reduce its thermodynamic efficiencies. It therefore becomes necessary to design high-performance equipment that is efficient not only in terms of energy quantity but also in the conservation of energy quality called exergy (Kotas 1985; Ofori-Boateng et al. 2012).

Because energy and exergy are directly related to cost and environment, it becomes necessary to assess thermodynamic feasibility of production technologies especially energy-intensive ones like thermochemical conversion technologies and use the results to pinpoint locations of inefficiencies for sustainable manufacturing. This can be done using sustainability assessment tools such as exergy or second law analysis, exergetic life cycle analysis (ExLCA) or exergo-environmental sustainability analysis, emergy analysis, thermo-economic analysis (or exergo-economic analysis) to mention only a few (Ofori-Boateng 2022; Lee and Ofori-Boateng 2013).

The exergy analysis tool has been used to estimate thermodynamic inefficiencies in biofuels production technologies especially biochemical and chemical conversion processes. For instance, Ofori-Boateng (2022) assessed the chemical exergy destruction of a biogas digester with algal biomass as feedstock and concluded that the algae biomass with chemical exergy of 6474 MJ was upgraded in quality by about sixfold when the digestate was recycled or used as biofertilizer or nutrients for algaculture. Other reports have indicated that recycling of waste generated in a manufacturing unit helps reduce the overall exergy destruction of the process (Iribarren et al. 2014; Ofori-Boateng and Lee 2014).

This chapter analyzes and reviews the degree to which resources are degraded during thermochemical waste conversion processes for the generation of bio-oil, syngas, biochar, bioelectricity and biochemicals.

[1] Entropy of a system is the measure of the quantity of unavailable energy to do work by the system. Entropy leads to the dissipation of wastes and emissions into the environment by a system or object.

## 5.2 Thermo-Environmental Sustainability and Assessment Tools

Due to population growth coupled with advanced technology, anthropogenic activities have implicated lives on the ecosystem in such a way that extra attempt must be made to offset the negative impacts at the same time provide better goods and services which are not detrimental to living organisms on earth. Until humans change their attitudes and reverse certain dangerous technological actions like the use of harsh chemicals (e.g., chlorofluorocarbons to manufacture certain products), the ecosystem will continue to be affected negatively. Through the struggles for creating peaceful and clean atmosphere for mankind, sustainability and sustainable development evolved. Sustainability has been defined in many ways, of which they are geared toward the effective use of resources to meet the needs of mankind today without compromising tomorrow's peace. Sustainability and Sustainable development have been explained in detail in Lee and Ofori-Boateng (2013). This chapter discusses the thermodynamic and environmental sustainability of thermochemical conversion technologies.

Currently, the aim of environmental safety is slowly drifting toward energy efficiency in industrial processing. In terms of environmental safety and thermodynamic efficiency with the use of thermochemical waste conversion technologies to produce value-added bioproducts, a sustainable production may not necessarily mean 'no waste' or 'no emissions' but rather involves the efficient utilization of eco-friendly raw materials and waste in the environment to produce cost-effective products through technologies that would not jeopardize the future (i.e., with minimal emissions). In a thermo-environmentally sustainable system, there is conservation of materials and energy (resources) whilst waste is reused or recycled. Again, the products generated are less polluting to the environment, and are economically viable and of high quality.

Sustainability analysis of a thermochemical conversion process is usually carried out with the aid of computer software (e.g. Aspen Plus, HYSIS, Pro II, etc.) to estimate the degree of thermodynamic inefficiencies (Doherty et al. 2013).

### *5.2.1 Exergy Analysis*

Exergy is the most valuable form of energy in a resource that has the potential to perform work. This implies that in any system, higher exergy efficiencies mean that there is small amount of exergy that is destroyed (Ofori-Boateng 2022). Through reversible processes, the useful part of energy extracted from a system is brought into thermal, mechanical, or chemical equilibrium (Rivero et al. 1999). This implies that in real life, the conversion of resources to useful products results in heat, work, and materials generation, and these products possess quality energy which will always be less than the raw materials or input resources of a system due to exergy degradation (Ofori-Boateng et al. 2012). The degree of entropy generation leading to the intensity of exergy destruction in a system or resource is the major cause of change in the exergy content of the final product (Kotas 1985).

The second law of thermodynamics (i.e. entropy and exergy) addresses the limitations of the first law (i.e. energy conservation) where the part of quality energy destroyed in a process is accounted for. Sustainability largely depends on how a process proceeds in the reversible way because production processes are irreversible due of energy dissipation and emissions into the atmosphere (Ofori-Boateng 2022).

Exergy analysis is a thermodynamic analysis tool which is used to investigate and detect points and units in a process or system where there is resource degradation, and further provide improvement options for efficient or sustainable production (Ofori-Boateng et al. 2012). Exergy analysis reveals the actual magnitude of thermodynamic inefficiencies of industrial processes, the main causes of these inefficiencies, and the cost of obtaining their internal flows and production to be assigned in a more comprehensive way (Valero et al. 1986; De Meester et al. 2006; Teles dos Santos and Park 2009). Lee and Ofori-Boateng (2012) and Ofori-Boateng (2022) have detailed the process of carrying out exergy analysis of energy systems to estimate thermodynamic imperfections and have suggested improvement options for biofuel production.

Depending on the system of interest, exergy consists of physical ($Ex_{ph}$), chemical ($Ex_{ch}$), kinetic ($c^2/2$),[2] potential (gx),[3] and nuclear ($Ex_{nu}$) exergies (Szargut 2005; Cornelisse 1997). For thermochemical conversion technologies, exergy analysis can be estimated considering both physical and chemical exergies.

*Physical exergy* is the maximum work that can be attained by a system when it changes from the initial state to the dead state. *Physical exergy* is the exergy due to a system's pressure and temperature which is mathematically defined by Eq. 5.1 (Szargut et al. 1988):

$$Ex_{ph} = (H - H_0) - T_0(S - S_0) \tag{5.1}$$

where $(H - H_0)$ = enthalpy change where '$_0$' indicates the dead state

$(S - S_0)$ = entropy change of the system where '$_0$' indicates the dead state

$T_0$ = environmental temperature = 298.15 K

$P_0$ = atmospheric pressure = 1 atm

The *physical exergy* of a perfect gas system is defined by Eq. 5.2 whilst that for perfect solid and liquid systems is represented by Eq. 5.3 (Szargut et al. 1988):

$$Ex_{ph} = c_p\left\{(T - T_0) - T_0 ln\frac{T}{T_0}\right\} + RT_0 ln\frac{P}{P_0} \tag{5.2}$$

$$Ex_{ph} = \left\{(T - T_0) - T_0 ln\frac{T}{T_0}\right\} - v_m(P - P_0) \tag{5.3}$$

where $Ex_{ph}$ = physical exergy,

---

[2] $c$ is the velocity relative to Earth's surface.

[3] $g$ is the constant of gravitational acceleration and $x$ is the height.

$c_p$ = specific heat capacity,
$v_m$ = specific volume at temperature $T_0$ (298.15 K),
$T$ = temperature of the system,
$R$ = universal gas constant (8.314 J mol$^{-1}$ K$^{-1}$),
$P$ = system's operating pressure, and
$P_0$ = initial pressure (1 atm).

*Chemical Exergy* of a system is the maximum net useful work obtained when a pure substance or working fluid goes into complete thermodynamic equilibrium with its environment of known chemical composition upon interactions in a system. Mathematically, chemical exergy is defined by Eq. 5.4 (Szargut et al. 1988):

$$\mathrm{Ex}_{\mathrm{ch},i} = \Delta G_f^0 + \sum_k N_k \mathrm{Ex}^0_{\mathrm{ch},k} \tag{5.4}$$

where $\mathrm{Ex}_{\mathrm{ch},i}$ = chemical exergy of species $i$
$N_k$ = number of moles of the $k$th reference species
$\Delta G_f^0$ = Gibb's free energy of formation of the reference reaction and
$\mathrm{Ex}^0_{\mathrm{ch},k}$ = standard chemical exergy of the $k$th reference species.

Exergy (i.e. the quality of energy within a product) is destroyed in a process due to entropy generation resulting in *internal or endogenous exergy destruction* (Rivero et al. 1999). In thermochemical conversion of waste, exergy destructions can result from inefficient operations of equipment as well as the quality of resources that enter (waste materials) and leave the production plant. The exergy of the waste, catalysts, reaction media, and all other input resources as well as the product gases that exit the facilities as wastes and emissions contribute to *external* or *exogenous exergy destruction* of the refinery (Rivero et al. 1999). The total exergy destruction of a thermochemical waste conversion facility can be calculated as the *sum of the internal and external exergy destructions* which is mathematically defined by Eq. 5.5 (Ayres and Ayres 1998; Cornelisse 1997; Ofori-Boateng et al. 2012):

$$\begin{aligned}\sum \mathrm{Ex}_{\mathrm{heat}} - \sum \mathrm{Ex}_{\mathrm{work}} + \sum \mathrm{Ex}_{\mathrm{mass,in}} - \sum \mathrm{Ex}_{\mathrm{mass,out}} &= \mathrm{Ex}_{\mathrm{destruction}} \\ &= T_0 S_{\mathrm{generation}} = I\end{aligned} \tag{5.5a}$$

$$\begin{aligned}\sum \left(1 - \frac{T_0}{T}\right) Q - W + \left(\sum_i m_i \mathrm{Ex}_i\right)_{\mathrm{in}} - \left(\sum_i m_i \mathrm{Ex}_i\right)_{\mathrm{out}} &= \mathrm{Ex}_{\mathrm{destroyed}} \\ &= T_0 S_{\mathrm{generated}} = I\end{aligned} \tag{5.5b}$$

where $\mathrm{Ex}_{\mathrm{heat}} = \sum \left(1 - \frac{T_0}{T}\right) Q$ = exergy destruction due to heat.
Q = heat duty of the system and
$T$ = the temperature of the stream/resource
$T_0$ = environmental/reference temperature (298.15 K).

$Ex_{work} = W$ = exergy of the system due to work production and can be calculated from the heat capacities and standard chemical exergies of the streams entering or leaving the system

$Ex_{mass,in} = \left(\sum_i m_i Ex_i\right)_{in}$ = material exergy input

$Ex_{mass,out} = \left(\sum_i m_i Ex_i\right)_{out}$ = material exergy output

$m_i$ = mass of $i$th component = mass of resource/stream

$Ex_i$ = exergy of $i$th component = exergy of resource/stream respectively

$Ex_{destruction}$ = exergy destruction

$S_{generation}$ = entropy generation

$I$ = irreversibility

In real situations, for every process, the exergy input always is higher than the exergy output due to irreversibility called exergy destruction $Ex_{destruction}$, which is a function of entropy generation (Ofori-Boateng et al. 2012).

Equation 5.6 mathematically defines *exergy efficiency*, which is the measure of a system's ability to transform raw materials into useful products (Rivero et al. 1999):

$$Ex_{efficiency} = \frac{Ex_{flow\text{-}out}}{Ex_{flow\text{-}in}} \tag{5.6}$$

## 5.3 Exergy Analysis of Thermochemical Conversion Technologies

The demonstration of process flow for some thermochemical conversion technologies in ASPEN Hysys/Plus software is presented in this section. Aspen Plus is a process modeling and simulation software used to evaluate chemical and physical systems (Peters et al. 2017) like thermochemical conversion technologies. Biomass or waste thermal decomposition using pyrolysis, combustion, gasification, or any other thermal treatment technologies can be analyzed in ASPEN Plus where the impacts of process conditions (e.g., temperature) on the yields and products compositions can be investigated based on their thermodynamic properties.

### *5.3.1 Exergy Analysis of Waste-to-Energy via Combustion and Incineration*

Figure 5.1 demonstrates the ASPEN Plus process flow diagram of waste combustion to energy, which depicts a basic combustion process model like most of the waste-to-energy combustion or incineration system discussed in this section.

From Fig. 5.1, the waste material (BIOMASS) is collected and transported to the combustion or incineration facility where it is dried in the DRYER before it enters the

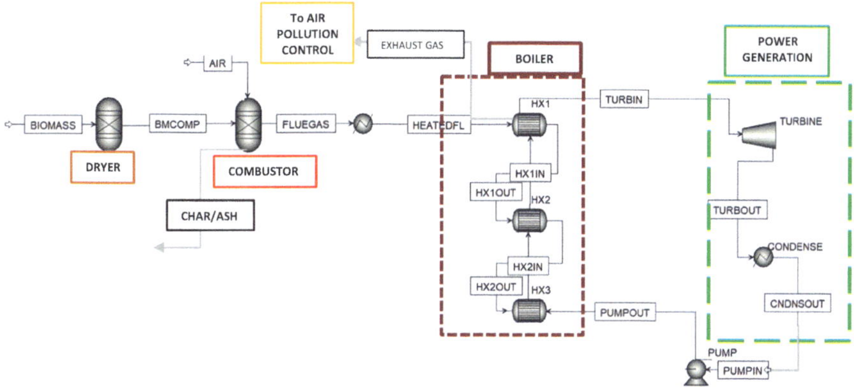


**Fig. 5.1** ASPEN Plus process flow diagram of waste combustion

COMBUSTOR as BMCOMP (dried biomass). Most incineration and combustion reactors may not need the waste to be crushed into smaller particle size before the reaction. Compressed air (AIR) transported by a blower enters the combustor at a pressure < $10^6$ kPa (~1.05 atm) and combustion of the waste occurs in the COMBUSTOR at 900–1200 °C where CHAR/ASH and FLUEGAS are generated. To control the oxygen molar content in the flue gas, the flow rate of the compressed air must be varied to correspond with the process temperature and inlet feedstock flow rate whilst considering the moisture content of the waste material. The FLUEGAS enters a heat exchanger where the temperature of the flue gas is elevated before it enters the boiler. The boiler comprises three heat exchangers arranged in parallel (HX1 vaporizer, HX2 superheater, and HX3 economizer). The tube side of the boiler can be set to 40–44 atm to generate steam with temperature of about 410–420 °C (Rispoli et al. 2021). The boiler generates steam (TURBIN) from the flue gas which is then sent to the steam turbine (TURBINE) for power generation using steam expansion. When the steam leaves the TURBINE as TURBOUT, it is condensed into CNDNSOUT by the condenser CONDENSE, and can be recycled back into the TURBINE or recirculated in the system.

Because most waste incineration facilities are not controlled with energy recovery due to their low performance, they can be integrated with natural gas-fired power plants in order to improve its energy and exergetic performance. In a hybrid system like this where superheated steam leaving the waste incineration boiler is used as fuel to elevate the temperature of water entering the natural gas-fired boiler, overall exergy efficiency and waste-to-energy efficiency were increased by about 0.03% and 0.84% respectively whilst the exergy destruction was reduced by about 1.74 MW minimizing the investment cost of the plant by 2.68% (Aghapour Sabagh et al. 2023). However, if the above hybrid system configuration is changed where the flue gas from the incineration unit is used to feed the intermediate-pressure (IP) turbine rather than the natural gas-fired boiler, the dynamics of exergy distribution changes and the overall exergy efficiency and waste-to-energy efficiency increases by 0.3%

and 7.4% respectively with exergy destruction being reduced by 2.2 MW compared to the natural gas-fired boiler configuration. This shows that for an exergy-efficient incineration system, equipment such as superheater (HX2) and economizer (HX3) can be removed from the boiler system to decrease the capital investment by about 4% (Aghapour Sabagh et al. 2023).

Vilardi and Verdone (2022) assessed the energy and exergy efficiencies of MSW incineration system after simulation in ASPEN Plus software. They reported that oxygen enrichment air helped increase the exergy efficiency of the incineration system whilst flue gas recirculation also increased the exergy efficiency by about 3% (i.e., from 31.1% to 34%) which is slightly higher than the exergy efficiency of incineration of MSW from South Beijing (15.59%) with waste recycling such as composting (Zhou et al. 2011). Similarly, the incineration of MSW from the city of Santo André in São Paulo State, Brazil, with energy recovery also recorded exergy efficiency of about 91% (Trindade et al. 2018). From the exergy analysis results, steam turbine's exergy destruction can be reduced by about 28% when technological improvements (e.g., turbine blade) are made which could make way for isentropic efficiency[4] to be increased (Trindade et al. 2018).

MSW sludge direct incineration with carbon capture (DI-CC) system for electricity generation is reported to have very low energy and exergy efficiencies (about 13.97% and 10.88% respectively) compared to a steam gasification-chemical loop combustion (SG-CLC) system for electricity generation (15.52% and 11.09% energy and exergy efficiencies respectively) (Shi et al. 2023). DI-CC system showed lower exergy efficiency because of the nature of feedstock (sludge). Gas combustion is reported to have compete oxidation compared to solid sludge incineration. An incinerator or combustor alone had a low exergy efficiency of 40–45% compared to steam gasifier (80–84%) (Shi et al. 2023).

Similar results were reported by Rispoli et al. (2021) who also assessed an oxygen-enriched MSW incineration waste-to-energy plant for the conversion of carbon dioxide ($CO_2$) to methane ($CH_4$) in terms of exergy and recorded an efficiency of 26.4% which is lower than incineration systems with oxygen enrichment (34% exergy efficiency) evaluated by Vilardi and Verdone (2022).

## 5.3.2 *Exergy Analysis of Waste Gasification*

Figure 5.2 represents a process model from ASPEN Plus showing the process flow diagram of waste gasification to syngas. Gasification of waste generates producer gas (or syngas) comprising $H_2$, $CO_2$, CO, and $CH_4$ with some traces of pollutants. All waste gasification processes reviewed in this section via exergy analyses mimic the

---

[4] Isentropic efficiency (defined for turbines, compressors, nozzles, etc.) is a measure of how well a turbine for instance can convert the energy in a fluid (e.g. gas) into mechanical power. It is defined as the ratio of the actual work output of the turbine to the maximum work output that could be attained under ideal isentropic conditions. When isentropic efficiency is high, the turbine is said to be more efficient (Skabelund et al. 2023).

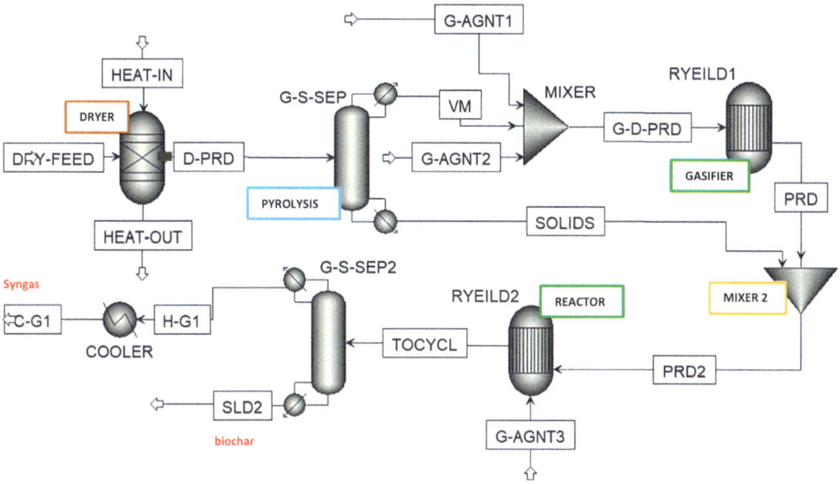


**Fig. 5.2** Aspen Hysys process flow diagram of waste gasification

flow diagram in Fig. 5.2. Before simulation in ASPEN Plus software for thermodynamic properties like enthalpy and entropy, energy and material balance calculations need to be carried out. Each unit operation in the gasification system must be assigned process conditions.

Gasification of waste into syngas comprises a series of reactions as seen in Fig. 5.2. First, water is removed from the waste material (DRY-FEED) by drying at 100–200 °C to obtain dried waste material (D-PRD) which then enters the reactor 1 (G-S-SEP) where it is heated at high temperatures (200–500 °C) to break the organic matter components down into gases (VM), tar and char (SOLIDS). The resulting product gases (VM) enter a mixer (MIXER) where they mix with gasifying agents (G-AGNT 1 and G-AGNT2, which can be steam, $CO_{2,}$ etc.) and exit the mixer as G-D-PRD. The mixture of volatile matter of the waste and gasifying agents enters the yield reactor (RYEILD1) where molecular compositions and phase equilibrium are estimated under the specific temperature and pressure conditions when Gibb's free energy is reduced (AlNouss and McKay 2018). The RYEILD 1 and RYEILD 2 can also function as a two-phase separator where the ash/char is separated at the bottom and gases exit at the top. The products (PRD) from the gasifier are mixed with the solids from the pyrolysis reactor (G-S-SEP) in MIXER 2 to form PRD2 where further reactions and separations occur in RYEILD2 and G-S-SEP2 to finally obtain biochar (SLD2) and hot hydrogen gas (H-G1) and cooled gas (C-G1). Several appropriate equations of state such as non-random two liquid (NRTL), Peng-Robinson, etc. can be used to replicate the actual and non-polar species in ASPEN Plus software.

Energy and exergy analyses with simulations and modeling in ASPEN Plus software for gasification of biomass may have some assumptions which may include the following:

1. Non-dimensional, kinetic-free model

2. Feed preparation and combustion processes are temperature-sensitive and happen instantaneously.
3. Gasification of waste is stable with uniform distribution of temperature and pressure under equilibrium conditions.
4. Isothermal and adiabatic phase refrigeration are considered.
5. Biochar is mainly carbon, ash and waste materials are unconventional property set unless user defined with components and properties in ASPEN Plus software. Nitrogen and air are assumed to have inert characteristics.
6. Heavy hydrocarbons like tar are considered unconventional unless otherwise set as user defined.
7. Some reports consider 110 °C and 1 bar pressure as superheated conditions in the gasification reactor for the gasifying agents.

After simulation in ASPEN Plus software, the exergy analysis can then be done using the steps listed in the previous section (*Exergy Analysis*) and the entropy and enthalpy values obtained. After obtaining the physical exergy and chemical exergy of each component in every stream, the exergy efficiency and destructions can be found. Locations within the gasification system where improvements need to be made are easily spotted. Step-by-step procedures for calculating exergies of steam components are outlined by Ofori-Boateng et al. (2012).

Gasification of poplar waste for hydrogen production with major units comprising a gasifier, catalytic tar destruction unit, cold-wet gas cleaning unit, syngas conversion, and hydrogen purification units, was assessed exergetically by Iribarren et al. (2014). The gasification system recorded an exergy efficiency of 48% which was similar to coal gasification system for hydrogen production (Iribarren et al. 2014). For this same gasification system, about 52% of the total exergy in the input resources to the plant was destroyed during the process (Iribarren et al. 2014).

The gasification of invasive weed (*Parthenium hysterophorus*) under steam-$CO_2$ atmosphere was simulated in ASPEN Plus software for thermodynamic properties, and the impacts of the unit operations in producing hydrogen gas on resource efficiency were assessed using exergy analysis tool (Nivash et al. 2024). It was observed that based on the yields of syngas composition namely 20% CO, 16% $CO_2$, and 3.7% $CH_4$ at 800 °C, gasification was an efficient technology for hydrogen production (Nivash et al. 2024). Again, gradual increase in temperature from 300–1000 °C showed rise in hydrogen concentration from about 51% to 63% whilst $CO_2$ concentration declined from 24% to 13.5% (all percent concentrations are by volume) (Nivash et al. 2024). The gasification of invasive weed for hydrogen production based on the investigations by Nivash et al. (2024), showed that the technology is exergetically efficient with very low exergy loss of about 0.9%.

Municipal solid waste (MSW) gasification in fluidized bed reactor hybrid system with solid oxide fuel cell and waste-to-energy (WtE) recovery unit was assessed using exergy analysis tool (Xiao and Wu 2024), and the results indicated high energy and exergy efficiencies of 40–41.4% and 36–37% respectively. Exergy efficiency of the hybrid system was impacted by many factors such as fuel cell utilization factor,[5] current density, and temperature. When the utilization factor and temperature were increased whilst reducing the current density of the fuel cell for MSW hybrid pyrolysis-fuel cell system resulted in an increase in exergy efficiency from about 37% to ~ 40% (Xiao and Wu 2024).

Okolie et al. (2020) used ASPEN Plus software to model gasification of soybean straw and flax straw in supercritical and subcritical water conditions (300–500 °C, 1:5 and 1:10 waste to water ratio, 0.13–0.8 mm particle size and 30–60 min residence time) into hydrogen gas and assessed the system's efficiency in terms of exergy. They recorded the maximum hydrogen production yield (about 44.4% of the total gas generated) at 500 °C, 1:10 waste-to-water ratio, 0.13 mm particle size, and 45 min residence time. Due to temperature gradient and absence of agitation in the tubular batch reactor, the experimental results deviated slightly from equilibrium values in a thermodynamic model, but alkaline catalyst was found to increase the gas yield (Okolie et al. 2020) thus, hydrothermal gasification of waste can be seen to exhibit good exergy efficiency based on the results of thermodynamic modeling in ASPEN Plus software.

Plasma gasification, which is a new innovative thermochemical conversion technology was integrated with High Temperature Proton Exchange Membrane Fuel Cell (HT-PEMFC[6]) cogeneration[7] system and used to valorize food waste into syngas and biochar (Yang et al. 2024). In a system like this, exergy analysis carried out by Yang et al. (2024) revealed that the plasma gasifier and the gas stack recorded the highest exergy destruction of about 62.1% and 32.2% of the total system's exergy destruction respectively with high gasification temperature (1000–1300 K) being the main contributing factor to exergy destruction.

Gasification reactor type and temperature are found to significantly impact exergy efficiency of the whole gasification process. For instance, gasifiers that utilize air as gasifying agent at high temperatures usually have higher exergy efficiencies compared to gasifiers that utilize steam and oxygen (Xiang et al. 2021). In their findings, Xiang et al. 2021 found that at a high gasification temperature of 1000 °C, the calorific value of the gas generated, and exergy efficiency are the highest. Gasification of MSW in an updraft fixed bed gasifier with air as gasifying agent is exergetically efficient (50%) than downdraft fixed bed gasifier (41.23%) due to difference in the reaction principles of reach reactor. For instance, the gas produced in the gasification zone of the updraft gasifier from the pyrolysis zone is fully utilized before leaving the furnace compared to downdraft gasifier where the combustible gas generation in the gasification zone is directly released from the furnace without utilization or further reaction which makes the gas composition constant hence higher calorific value (Xiang et al. 2021).

[5] Fuel Utilization factor is the amount of initial fissile material needed per unit energy and it is proportional to the current density when the fuel is consumed (Dincer and Zamfirescu 2014).

[6] HT-PEMFC (High Temperature Proton Exchange Membrane Fuel Cell) also called High Temperature Polymer Electrolyte Membrane Fuel cell is a type of fuel cell which with fast reaction kinetics hence high energy efficiency, and it allows for cogeneration of heat and power.

[7] A cogeneration system is also called combined heat and power (CHP) plant which is used to produce heat and electricity simultaneously in the same unit.

The combustion section of a supercritical water gasification (SCWG) of MSW sludge into syngas and biochar having gas turbine attached for energy generation was found to be exergetically efficient (70.78%) compared to that of waste-to-energy incineration or combustion system (41.78%) (Shi et al. 2023).

The exergetic performance of plastic waste integrated gasification-combined cycle with heat recovery steam generator (HRSG), and another with gas heater (GH) for electricity generation were evaluated with conclusion that the location and ways of using recovered heat are important for better exergy efficiency (Lee and Im 2024). For the gasification system with gas heater, the efficiency of the power cycle increased by about 8.16% compared to the gasification with heat recovery steam generation because of the heat used to preheat the air supplied to the gasifier in the GH instead of being used solely to generate steam in the HRSG. Nonetheless, the quality of energy destruction in the combustion chambers for both configurations was low (i.e., 140,201 kW and 135,630 kW for HRSG and GH respectively) (Lee and Im 2024).

Bhattacharya et al. (2014) investigated the effect of oxygen concentration in a gasifying agent on exergy efficiency during rice straw gasification (with air separation unit) for hydrogen production. Their results indicated that with 95% oxygen in the gasifying agent and 4.0 gasifier equivalence ratio[8] (ER), about 107.8 g of hydrogen per kg of rice straw was generated at cold gas efficiency of 70% and exergy efficiency of 67%. In the gasifier, about 27% of the total exergy of the rice straw and gasifying agent was degraded during the combustion process (Bhattacharya et al. 2014).

Depending on the gasification conditions, exergy efficiency of waste gasification is found to range from 40–75% with main irreversibilities coming from the combustion reactors. Waste recycling drastically decreases the exergy destruction related to waste gasification facilities.

### 5.3.3 *Exergy Analysis of Waste Pyrolysis*

Figure 5.3 shows a process flow model of waste pyrolysis into syngas, bio-oil, and biochar as simulated in ASPEN Plus software. The procedure for exergy calculations is the same for all the technologies discussed in this chapter. For process improvement, ASPEN Plus model can incorporate kinetic reaction models of pyrolysis of waste based on the main compositions and characteristics of the waste materials used to predict products yields.

The pyrolysis of waste into oil, char, and gases can be grouped into 3 units namely (1) Waste preparation, (2) Pyrolysis Unit, and (3) Products separation Unit. From Fig. 5.3, in the biomass preparation unit, the waste material is first dried (DRYER) at 120–150 °C and 1 bar to reduce the moisture content to at least 4–15%, sent to the

[8] Equivalence ratio (ER) is a critical product quality parameter in gasification which is the ratio of oxygen or air concentration fed to a gasifier to the oxygen amount required to bring the combustion reaction to a completion.

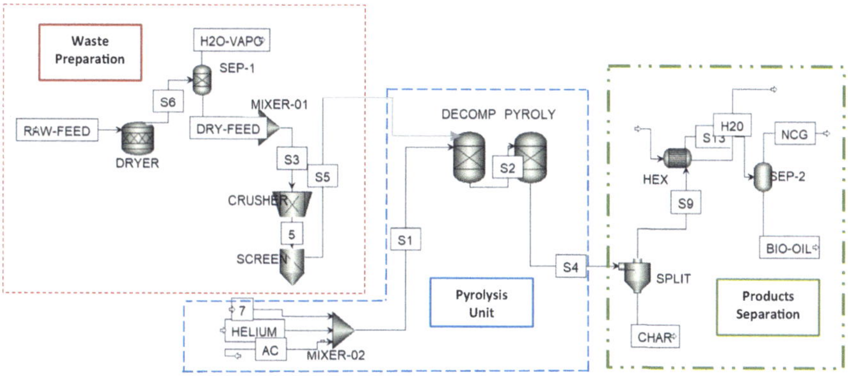


**Fig. 5.3** ASPEN Plus process flow diagram of waste pyrolysis

separator SEP-1 to split the water vapor (H2O-VAPO) from the dried waste (DRY-FEED) which is then crushed into pieces in CRUSHER to increase its surface area for efficient heat transfer during pyrolysis. The dried and crushed waste material (S5) enters the pyrolysis reactors (DECOMP and PYROLY). The RYIELD DECOMP reactor decomposes the organic matter content of the waste at 1 bar and up to 550 °C before it enters the PYROLY reactor where further decomposition occurs at elevated temperatures of 550–700 °C to generate condensable gases, biochar, and other gases (S4). The product S4 is then split into char (CHAR) and gases (S9) in a cyclone (SPLIT). The condensable gases in product stream S9 are cooled to condense in a heat exchanger (HEX) and all products from this unit (S13) enter a separator (SEP-2) where the separation of the condensed condensable gases (pyrolysis oil BIO-OIL) and non-condensable gases (NCG) is separated.

Considering energy and exergy balance calculations with physical and chemical exergies of each stream calculated, the exergy efficiency for the process outlined in Fig. 5.3 can be found. Exergy analysis of pyrolysis of waste in literature discussed in this section basically follows the ASPEN Plus model of pyrolysis depicted in Fig. 5.3.

An ASPEN Plus model and exergy analysis for flax straw pyrolysis was carried out by Temireyeva et al. (2024) who concluded that slow pyrolysis of flax straw (at 400 °C) exhibited lower products yields (42.8% char, 35.8% oil, and 21.4% gas) and lower exergy efficiency (89%) compared to fast pyrolysis (at 500–600 °C) with high oil yield (40.3% and 29.5% char) and exergy efficiency of about 98%. With total input energy of 0.994 MJ (at 400 °C), 1.009 MJ (at 500 °C) and 1.016 MJ (at 600 °C) for fast pyrolysis of flax straw, the total energy outputs were 0.296 MJ biochar, 0.429 MJ pyrolysis oil, and 0.224 MJ biogas at 600 °C which were slightly lower than the exergy contents of the products (0.313 MJ biochar, 0.462 MJ bio-oil, and 0.244 MJ biogas) (Temireyeva et al. 2024) implying that the waste (flax straw) which entered the pyrolysis unit was upgraded into products with high quality or useful energy. The overall exergy efficiency recorded by Temireyeva et al. (2024) was about 86.9–94.8% for slow pyrolysis and 88.9–98% for fast pyrolysis. Exergy efficiency of waste pyrolysis is higher compared to other biomass conversion technologies like transesterification of algal biomass into biodiesel with exergy efficiency of 86–89% (Ochoa-Garcia et al. 2018; Peralta-Ruiz et al. 2018).

Thermodynamic sustainability assessment of microwave pyrolysis of horse manure was evaluated by Hosseinzadeh-Bandbafha et al. (2022) at optimal temperature of 350 °C and 2:1 biomass-to-catalyst ratio. During microwave-assisted pyrolysis, high reaction temperature significantly affected the exergy efficiency of the system, and the exergy destruction due to irreversibilities and energy dissipation was more than 50% of the total exergy of input resources. However, at optimum conditions of 350 °C and 1:2 catalyst-to-biomass ratio, the recorded exergy efficiency of the system was about 49.2% (Hosseinzadeh-Bandbafha et al. 2022). Horse manure possesses high exergy content of about 297–595 kJ which was not impacted during the heating process but was found to decrease upon mixing with activated carbon catalyst. This implies that using more resources in microwave pyrolysis would increase exergy destruction. Again, the highest exergy content of the syngas generated (12.796 MJ/kg) was attained at 550 °C and 2:1 catalyst to horse manure ratio whilst pyrolysis oil exergy was highest at 450 °C (23.235 MJ/kg) & 550 °C (22.554 MJ/kg) and 1:2 & 1:1 catalyst-to-biomass ratio respectively (Hosseinzadeh-Bandbafha et al. 2022). The biochar obtained at similar conditions contained exergies (i.e., 13.704 MJ/kg at 550° C and 2:1 catalyst-to-waste ratio; 13.199 MJ/kg at 450° C and 1:2 catalyst-to-waste ratio and 14.986 MJ/kg at 350° C and 1:1 catalyst-to-waste ratio) slightly higher than those for syngas (Hosseinzadeh-Bandbafha et al. 2022). Thus, if the targeted product during microwave-assisted pyrolysis of horse manure is bio-oil, the process conditions must be set at 1:2 catalyst-to-waste ratio with heating no more than 450 °C.

Thermochemical conversion technologies which have solar hybrid systems for heat electricity recovery are reported to increase energy and exergy efficiencies hence reducing the systems' overall operational cost (Chen et al. 2022). A novel pyrolysis technology which incorporates solar char-cycling (SCCP[9]) for converting raw biomass into biochar was modeled in ASPEN Plus software and reported to have zero combustion $CO_2$ emissions leading to high exergy efficiency of about 76.51% (with an improved gas of higher heating value (HHV) of 13.66 MJ/kg) compared to 68.49% for solar direct pyrolysis (SDP[10]) and 72.43% for autothermal pyrolysis (AP[11]) (Li et al. 2024). For the SCCP system, with biomass exergy of 106.9 kW that entered the unit, about 19.62 kW exergy of syngas, 58.98 kW exergy of pyrolysis oil, and 15.43 kW exergy of biochar were generated with about 12% of the exergy content in the waste material being lost during pyrolysis. These exergy values of the products are higher for SCCP system compared to SDP and AP at the same conditions (Li et al. 2024). Thus, solar hybrid system with pyrolysis of waste is advantageous

[9] Solar char cycling pyrolysis (SCCP) is a method of pyrolysis where high-temperature process heat is supplied by both solar and biochar either from outside or generated within the pyrolysis unit.

[10] Solar Direct Pyrolysis (SDP) is a method of pyrolysis where high-temperature process heat is supplied by solar energy.

[11] Autothermal pyrolysis is a method of pyrolysis where only air is used as the fluidizing gas thus the energy supply for pyrolysis is provided by partial oxidation of the pyrolysis products in the reactor.

to direct pyrolysis and autothermal pyrolysis as the SCCP systems overcome the window contamination problem seen with AP and SDP (Li et al. 2024).

Particle size of waste material, reaction time, and temperature during pyrolysis is found to affect the exergy efficiency of the system. For instance, the highest energy (72.9%) and exergy efficiencies (68.4%) of fast pyrolysis of municipal green waste in an auger reactor were recorded at 550 °C, 3 min residence time, and 2 mm particle size (Hasan et al. 2024). When the retention time and particle sizes were increased to 4 min and 10 mm respectively at 550 °C, the exergy efficiency decreased to a low 43.5%. Thus, though large particle size can increase char formation, it can reduce pyrolysis temperature, leading to incomplete combustion, and can drastically affect the overall exergy efficiency of fast pyrolysis (Hasan et al. 2024).

Similarly, fast pyrolysis of hybrid poplar woodchips modeled in ASPEN Plus indicated overall exergy efficiency of 71.2% with the combustor and pyrolysis reactor recording the highest exergy destruction (Peters et al. 2014). The performance of the fast pyrolysis plant was improved with slightly higher exergy efficiency of 73.2% when the hot exhaust gases from the combustor are directly introduced into the dryer without using any of it for preheating the combustion air. This strategy reduces the inefficiencies in the dryer and heat exchanges due to material recycling in the plant hence improving the overall performance of the plant (Peters et al. 2014).

With similar reactor's operation temperature of 400–600 °C, wood chips pyrolysis simulated in ASPEN Plus software showed reactor exergy efficiency of 66–77% (Dhaundiyal and Atsu 2021) which is in close agreement with results of Hasan et al. (2024) and Peters et al. (2014).

Table 5.1 summarizes the exergy analysis results of the pyrolysis of different kinds of waste materials into value-added bioproducts.

**Table 5.1** Exergy efficiencies of different methods of Pyrolysis of waste materials (Hosseinzadeh-Bandbafha et al. 2022)

| Pyrolysis Technology | Type of waste material | Pyrolysis conditions | Exergy efficiency, % |
|---|---|---|---|
| Solar-microwave hybrid pyrolysis | Rice straw | 700 W microwave power, 5 wt% of waste as additive to biochar, 20 min residence time | 56.7 |
| Conventional | Rice husks | 1200 °C | 65.27 |
| Conventional | Wheat straw | 400 °C, 0.2 MPa, 100 s, nitrogen as carrier gas | 60.62 |
| Conventional | Sewage sludge | 530 °C, 5.7 min | 73.6 |
| Conventional | Torrefied sewage sludge | 275 °C for 24 min | 89.9 |
| Conventional | Flax shives | 500 °C, 5 min, nitrogen as carrier gas | 91.05 |

### 5.3.4 Exergy Analysis of Waste Torrefaction

Figure 5.4 shows a process flow model of waste torrefaction into syngas and torrefied biomass as simulated in ASPEN Plus software. The dynamics of exergy distribution within the torrefaction facility as shown in Fig. 5.3 may differ depending on flue gas recycling or not. During ASPEN Plus modeling, torrefaction can be considered as a devolatilization process occurring in a range of temperatures (150–300 °C) where the gas stream (F-GAS, which is recycled) is mixed with the waste material (RAWPELL) in the RStoic reactor (D1) for torrefaction to begin. The content of D1 is then sent to a separator (F1) where the exhaust gas (EX-GAS) is separated from the slightly torrefied biomass (DRI-PELL). The solids DRI-PELL then move into the reactor (T1) and mixes with gas stream (F-GAS2, recycled) for further torrefaction. The product from TI enters the separator (F2) which splits it into gas product (GASPROD1) and intermediate product (INTER). Torrefaction is further carried out in reactor (T2) using recycled gas stream (F-GAS3) and the products are finally separated by F3 into the torrefied biomass (TORRPELL) and gas products (GASPROD2).

Temperature is found to affect irreversibilities in torrefaction facilities. For instance, torrefaction of pigeon pea stalk and eucalyptus in a tubular reactor at 200–300 °C with constant residence time and heating rate under nitrogen environment recorded high exergy efficiency at 250 °C where the solid yield for both waste materials were 63–64% (by-product liquid yield of 19–29% and 8–18% non-condensable gases (NCG)) (Singh et al. 2020). The exergy of the product solid for pigeon pea stalk at 250 °C was 13.66 MJ/kg waste which is higher than that for torrefied oil palm fiber (12.81 MJ/kg) and torrefied rice husks (12.07 MJ/kg) but lower than torrefied eucalyptus (14.05 MJ/kg), torrefied sugarcane bagasse (15.07 MJ/kg), and torrefied sawdust (16.38 MJ/kg) (Singh et al. 2020; Granados et al. 2014). Operating temperature for torrefaction of loblolly pine in the range of 200–260 °C is reported to be favorable for exergetically efficient systems and products (Yan et al. 2010). Thus, severe torrefaction temperature of 270–300 °C is not advantageous to agricultural residues torrefaction in terms of energy and exergy efficiencies.

Due to oxygen: carbon (O/C) and hydrogen: carbon (H/C) ratios of rice husk, torrefaction at 250–350 °C for 1 h improved the chemical exergy content of the

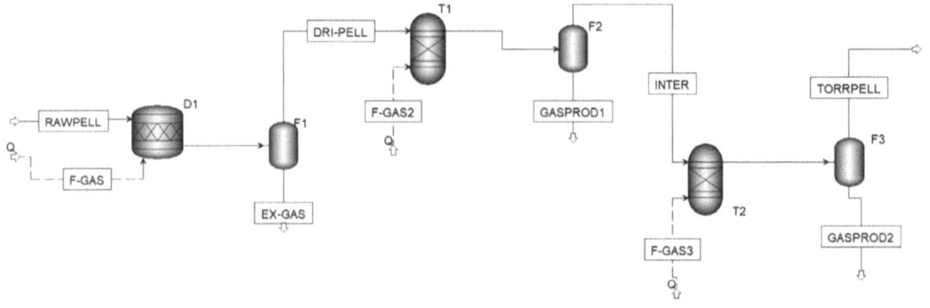


**Fig. 5.4** ASPEN Plus process flow diagram of waste torrefaction

torrefied rice husks with an overall exergy efficient of torrefaction process being 21–30% (Manatura et al. 2017). However, it is interesting to know from this investigation that gasification of high exergy content torrefied rice husks in a bubbling fluidized bed gasifier into syngas destroyed part of the quality energy in the feedstock hence reducing exergy efficiency (Manatura et al. 2017).

To evaluate the effects of torrefaction on gasification performance via energy and exergy analyses, microwave torrefaction (MT) and conventional torrefaction (CT) of herb residues were modeled in ASPEN Plus and validated by experimental data by Jiao et al. (2022). Their results indicated that microwave torrefaction improved the exergy content of herb residue by about 17.21% (17.68 MJ/kg exergy of torrefied herb residues) with exergy contents of the generated gas product and tar being 7.93 MJ/kg and 3.13 MJ/kg respectively (Jiao et al., 2022).

An integration of steam gasification with torrefaction can enhance the performance of the thermochemical conversion process compared to single direct gasification or torrefaction. For instance, Zhao et al. (2024) carried out exergy analysis of an integrated system which used torrefied food waste (5 tons/h) as gasification feed to produce hydrogen and found out that direct gasification of the waste had lower exergy efficiency (57.2%) compared to the integrated system (59.6%).

## 5.4 LCA of Thermochemical Waste Conversion Technologies

### *5.4.1 Life Cycle Assessment (LCA)*

Life cycle analysis (LCA) is an environmental assessment tool which is used to evaluate the impacts of products over their life cycle (from raw material extraction to product consumption). Even though biofuels combustion releases minimal emissions, the technologies employed to produce biofuels use energy and fuels sourced from fossils, and this is a major concern to the biofuel industry. Usually, net energy and material balance calculations are included in LCA results for production technologies such as pyrolysis, gasification, etc. The main aim of LCA is to compare the range of environmental burdens assignable to products and processes to select the best one with ease (Lee and Ofori-Boateng 2013). Figure 5.5 summarizes the methodology for LCA of bioenergy and biofuels production units.

The LCA methodology as depicted in Fig. 5.5 is a standardized method for environmental impact assessment over the lifespan of a process, a system, or a product (Tsangas et al. 2023). LCA methodology starts with the definition of goal and scope where the functional unit and the system boundaries are set. Commonly used life cycle boundaries include '*cradle-to-grave*' (from feedstock generation to product disposal), '*cradle-to-gate*' (from feedstock generation to production), etc. for waste materials conversion into energy. '*Gate-to-grave*' for instance may be suitable when the process starts from waste collection and sorting all through to consumption and

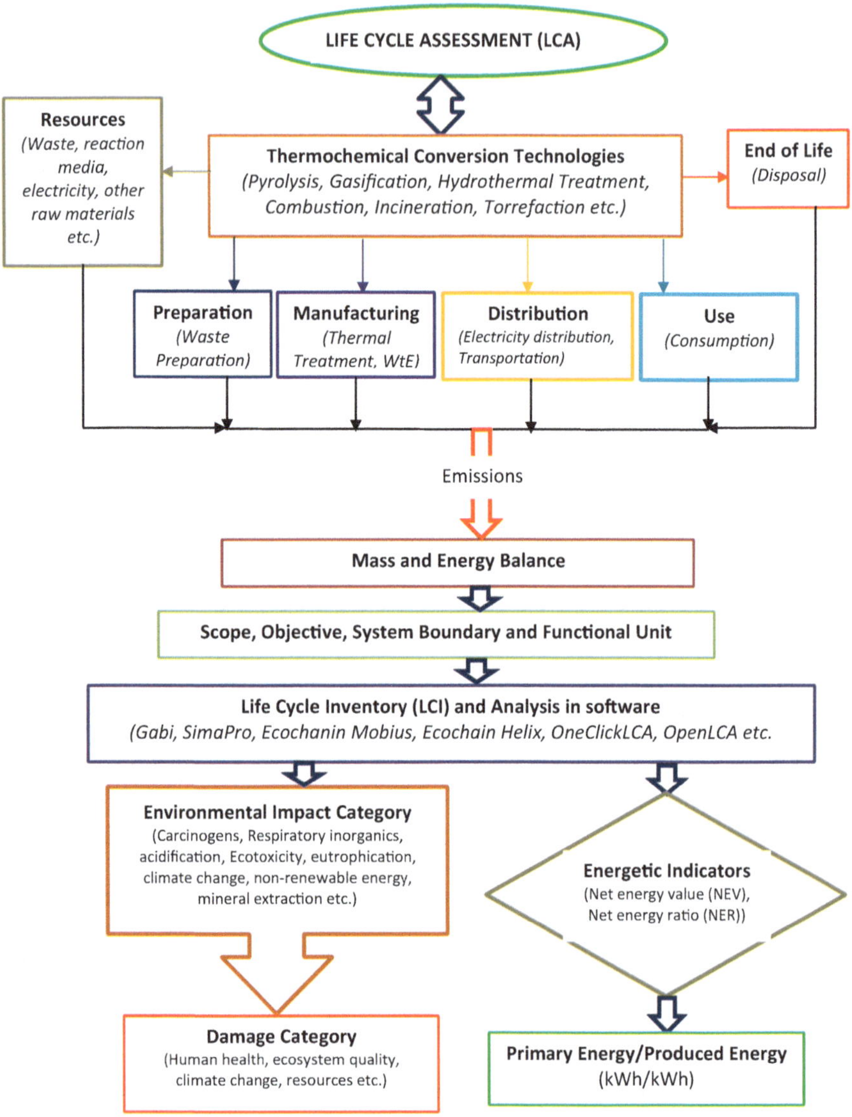


**Fig. 5.5** LCA methodology

disposal. *Functional unit* is a reference unit for the measurement of environmental performance of a system, a product, or a process. The *life cycle inventory (LCI) phase* is when data collection is spelt out for *life cycle impact assessment (LCIA)* to be carried out to address all the impact categories, and results are interpreted at the life cycle interpretation phase. The LCIA phase establishes a link between the process (or product) with its potential environmental impacts. Finally, conclusion and recommendations are made based on ISO standards.

Environmental impact assessment via LCA usually involves life cycle inventory (LCI) (see Fig. 5.5) which basically relates the potential impacts to the environment, human health, and resources depending on set characterization factors (Andreasi Bassi et al. 2023). According to Mulya et al. (2022), LCA characterization methods for waste conversion include CML (Chain Management by Life cycle assessment, which uses the $kgCO_{2eq}$ for global warming potential, for instance), TRACI (Tool for the Reduction and Assessment of Chemical and Other Environmental Impacts, which uses same unit as CML), Eco-indicator 99 method, PEF (Product Environmental Footprint, which uses fossil $kgCO_{2eq}$ or biogenic $kgCO_{2eq}$, etc.), ReCiPe, etc. (Ofori-Boateng and Lee 2014; Neri et al. 2018; Li et al. 2010). Environmental impact methods have many kinds of impact categories and are selected based on system boundary of the process. Apart from the commonly used impact categories listed in Fig. 5.5, there are several others which include but are not limited to:

1. Ozone layer depletion (ODP)
2. Photochemical ozone creation (POCP)
3. Marine aquatic ecotoxicity (MAETP)
4. Human toxicity (HTP)
5. Freshwater aquatic ecotoxicity (FAETP)
6. Terrestrial ecotoxicity (TETP)
7. Depletion of abiotic resources (ADP) (fossil fuel depletion)

Climate change impacts (global warming potential (GWP) as the characterization factor) are mostly included in over 90% of reported research results on LCA of thermochemical conversion technologies.

### *5.4.2 LCA of Waste Gasification, Combustion, and Incineration Systems*

Incineration facilities are found to emit nitrous oxides ($N_2O$) through the heterogeneously catalyzed oxidation of char-bound N-species as well as gas-phase reaction of nitrogenous groups in the volatile compounds generated during incineration and combustion (Gutierrez et al. 2005). Incineration and combustion facilities generate bottom ash, fly ash, and leachates which are hazardous, but their contamination characteristics are not well known in literature. Han et al. (2024) carried out a research investigation on the occurrence and chlorinated paraffin contamination properties of the solid by-products from incineration and combustion. Their results showed high concentrations (448 ng/g fly ash, 124–7230 ng/g in bottom ash) of short- and long-chain chlorinated paraffins on the adsorbed phases compared to the dissolved phase of the solid by-products (predominantly $C_{10}Cl_{6\text{-}7}$ in fly ash and leachate, and $C_{13}Cl_{6\text{-}7}$ in bottom ash as well as the $C_{14}Cl_{7\text{-}8}$ group in all the solid products). Annually, chlorinated paraffins (organic pollutants) from all solid by-products generated in MSW incineration and combustion facilities are hazardous and they release toxic compounds in the range of 66–7510 kg (Han et al. 2024). The main contributors of emissions in thermochemical conversion processes are dryers and combustors as revealed from various studies discussed below for incineration, combustion, and gasification of waste materials.

LCA of poplar waste gasification into hydrogen gas revealed a promising energy performance with the largest environmental impacts coming from the biomass pretreatment and syngas cleaning units (Iribarren et al. 2014).

The LCA of microwave-assisted pyrolysis of horse manure was used as an environmental assessment tool (SimaPro software with Ecoinvent database) to determine the impacts of thermal processing of waste into value-added bioproducts (Hosseinzadeh-Bandbafha et al. 2022). The activated carbon catalyst recorded environmental impacts of 1.9 mPts/kg (milli-ecopoints per kg) whilst collection and transportation of horse manure contributed about 0.995 μPts/kg. Electricity utilization in the operation of the microwave-assisted pyrolysis facility contributed about 0.319 mPts/kWh to the total environmental impacts of about 182.53 mPts/kWh (Hosseinzadeh-Bandbafha et al. 2022).

LCA carried out on a MSW sludge supercritical water gasification system with gas turbine for power generation showed the highest global warming potential of 848.51 tons $CO_{2eq}$ per day with highest greenhouse gas (GHG) emissions compared to steam gasification of MSW sludge with chemical looping combustion (SG-CLC) system (162.28 tons of $CO_{2eq}$ per day) implying that SG-CLC system has a better performance in mitigating greenhouse gas emissions and can reduce carbon emissions by about 57.11% (Shi et al. 2023).

A comprehensive LCA of a plasma gasification (at 1200 °C) of excavated waste-to-hydrogen gas with integration of carbon capture was carried out by Aich et al. (2024) whose results indicated superior environmental performance compared to other waste-to-hydrogen processes. The global warming potential and terrestrial ecotoxicity impacts for instance were found to be respectively about 30% and 78% less for plasma gasification system with sorption enhanced/gas–water shift process compared to plasma gasification with gas–water shift/acid gas removal process although the impacts on resources were high for both systems plasma gasification system (Aich et al. 2024). Similarly, the technology of plasma gasification of waste materials which can simultaneously recover energy and materials was found to have more environmental benefits than incineration waste-to-energy facilities (Sanjaya and Abbas 2023).

Ayub et al. (2023) developed a plasma gasification-based tri-generation process for dimethyl ether (DME) and power generation to transform poultry litter into syngas and biochar where LCA results indicated positive environmental performance compared to coal-based DME system. The main contribution to the low score for the waste-based plasma gasification-DME production system was the bio-based syngas that was used as feedstock for DME generation (Ayub et al. 2023).

The gasification of mixed plastic waste to produce hydrogen and methanol was found to cause rise in GHG emissions by about 36% but low-level contributions from smog formation ($-4.69 \times 100$ kg $O_{3eq}$/kg waste-based hydrogen compared to 8 × 100 kg $O_{3eq}$/kg for fossil-based hydrogen), acidification (1.7 g $SO_{2eq}$/kg for waste-based methanol compared to about 5.5 g $SO_{2eq}$/kg for fossil-based methanol), eutrophication (−0.03 $gN_{eq}$/kg for waste-based hydrogen compared to about 3.2 $gN_{eq}$/kg for fossil-based hydrogen), particulates (−0.76 $PM2.5_{eq}$/kg for waste-based hydrogen compared to 1.9 $PM2.5_{eq}$/kg for fossil-based hydrogen) due to co-product credit of steam and electricity exports (Afzal et al. 2023). Without credits from co-products, the waste-based hydrogen system would have resulted in impacts similar to conventional hydrogen system, except for smog. Again, the use of natural gas contributed about 12–90% of total environmental impacts from the waste-based hydrogen system except for water use. Catalyst used in the waste-based hydrogen system rather contributed highest impact (7–22% of total impacts) of carcinogenic, eutrophication, ecotoxicity, and noncarcinogenic due to large amount of the bed material used per unit of product generated (Afzal et al. 2023). In a similar study, Arena et al. (2023) concluded from their LCA results that the best case performance for climate change environmental impact category is the plastic waste-to-hydrogen system with carbon capture followed by combustion with carbon capture.

Considering the cradle-to-grave approach of LCA, olive pomace waste gasification (using updraft fixed bed gasifier[12]) was found to have very low impacts for ozone layer depletion (2.74 × $10^{-6}$ kg CFC-11 $_{eq}$/20 kg olive pomace) and global warming potential (36.99 kg $CO_{2eq}$/20 kg olive pomace) only when the biochar and tar generated are utilized (Ozturk et al. 2023).

In a MSW combustion or incineration waste-to-energy system, there is avoidable exergy destruction due to the process and resource which can lead to drastic emission index reduction from 531 to −224 kg $CO_{2eq}$./MWh and energy displacement of about 3.69 kWh per unit available electric energy (Trindade et al. 2018). In a similar incineration system but hybridized with solar thermal system to process MSW for the recovery of heat and electricity, it was observed and concluded from an environmental impact assessment results that, the hybrid incineration system released more $CO_2$ emissions as the amount of MSW charged into the reactor increased, thus the quality or composition of MSW as well as temperature increase in the reactors directly influence $CO_2$ emissions (Wang et al. 2024).

Han et al. (2024) assessed the environmental impacts of MSW incineration system for energy recovery, and the LCA results showed that high plastics components in the MSW stream contributed high levels of GHG emissions every month ranging from 11.2–622.4 kg $CO_{2eq}$/ton MSW depending on the percentage of plastics. In order to reduce these emissions, MSW must be sorted at source to isolate plastic waste from other waste materials since the GHG emissions from only plastic waste combustion for energy recovery are minimal compared to when mixed with other types of waste (Afzal et al. 2023; Han et al. 2024).

Waste gasification systems are effective to minimize environmental impacts like acidification, nutrient enrichment, and photochemical ozone formation potentials although GHG emissions are relatively high compared to incineration systems with

[12] Updraft gasifiers were found to be more exergetically efficient than downdraft gasifiers (Xiang et al. 2021).

energy recovery (Tang et al. 2020). On the other hand, incineration systems for energy recovery are reported to be more energy efficient compared to gasification technologies with waste-to-energy components (Tang et al. 2020).

However, comparing coal-sewage sludge co-incineration and sewage sludge incineration systems, LCA results showed GHG emissions of 1133.33 kg $CO_{2eq}$/t dried sludge and 350.89 kg $CO_{2eq}$/ton dried sludge respectively, implying that co-incineration of waste and fossil-based coal is over 3 times more detrimental in terms of global warming and climate change compared to waste incineration (Yang et al. 2023). It is therefore environmentally beneficial to incinerate waste instead of depending fully on coal incineration for heat recovery which emits about 828.63 kg $CO_{2eq}$/ton coal, and the application of exhaust gas heat recovery systems to incinerators may offset up to about 8–23% of total GHG emissions from incineration facilities (Yang et al. 2023).

### *5.4.3 LCA of Waste Pyrolysis and Hydrothermal Liquefaction Systems*

Hydrothermal Liquefaction (HTL) is one of the promising technologies used to convert waste materials into liquid biofuels like diesel, gasoline, and biochemicals. LCA has been used in many ways to assess the environmental impacts of thermochemical conversion technologies, but few reports are available in literature for LCA of HTL though there are a significant number of HTL commercial scale facilities operating in most part of the world. In order to attain improvement options for emission reduction and ensure sustainable HTL process, environmental impact assessments via LCA tools may be employed.

LCA evaluation results on a HTL system for the conversion of agricultural waste corn straw into ethyl levulinate showed GHG emission reduction of about 37–90% compared to fossil diesel (Zhuo et al. 2023; Zoppi et al. 2023). An environmental impact assessment of an integrated HTL-aqueous phase reforming system for valorizing corn stover into biofuel is found to have a global warming potential of about 58.4 g $CO_{2eq}$/MJ biofuel with main burdens originating from the electrolysis step in the lignin-rich stream as well as the thermal duty and platinum use (Zoppi et al. 2023). Though catalytic HTL of cotton stalk resulted in higher yield of bio-oil compared to non-catalytic HTL, LCA results showed that it contributed the highest environmental impact (26.36–75.69% of overall impacts) as opposed to 1.95–7.23% for non-catalytic HTL (Lin et al. 2023). This shows that product yield and quality does not have a significant impact on the environmental benefits or performance of the whole process.

A 17.8 MW electricity generation plant utilizing pyrolysis oil produced from tire waste pyrolysis plant was evaluated by Tsangas et al. (2024) using LCA where the system boundary began from the shredding of waste tire, continuing from pyrolysis process through electricity generation to the management of the pyrolysis oil, biochar, pollutants, and other waste, all within the same facility. Using the CML 2001 impact characterization method and comparing results with light fuel oil production for electricity generation, it was deduced from the LCA results that tire pyrolysis oil production is more environmentally friendly compared to light fuel oil production (Tsangas et al. 2024).

The environmental impacts of co-pyrolysis of agricultural-forest residues (corncobs and wood waste) mixed with sewage sludge were compared with municipal sewage sludge pyrolysis via LCA and the results indicated that the single pyrolysis of sewage sludge caused increase in all 18 environmental burdens assessed, specifically global warming and ecotoxicity potentials (348.75 kg 1,4-DCB[13]/1000 kg biomass) but co-pyrolysis with agricultural residues drastically reduced the impacts by about 10–90% (Mohamed et al. 2024). Corncobs pyrolysis recorded the highest positive energy balance (+1368 kWh per 1000 kg of corncobs) and highest energy recovery which was about 59–181% of the total energy requirement of the process compared to -280 kWh per 1000 kg for single sewage sludge pyrolysis (Mohamed et al. 2024). The transportation of the waste materials from collection points to pyrolysis facility also contributed to terrestrial ecotoxicity due to the combustion of diesel fuel in the vehicle emitting $NO_x$, $SO_x$ and particulate matter into the environment (Mohamed et al. 2024). Electricity use in this pyrolysis unit using both sewage sludge and agricultural residues as feedstock contributed to marine ecotoxicity, human carcinogenic toxicity, and human noncarcinogenic toxicity. Mohamed et al. (2024) concluded that for a beneficial environmental impact of co-pyrolysis of waste, waste mixing ratios of 50:50 must be considered to achieve low global warming potential (468.95 kg $CO_{2\ eq}$/1000 kg). Also, sewage sludge single pyrolysis raises environmental concerns which may require more strategies to curb the burdens and make them more sustainable. Co-pyrolysis of sewage sludge with forest residues and other lignocellulosic materials is however found to reduce the overall environmental burdens by about 58–83% depending on the type of waste materials used as co-feed (Huang et al. 2023).

The environmental impacts of pyrolysis of food waste comprising mainly tomatoes, carrots, and cucumbers were investigated via life cycle assessment and the results indicated that, as the pyrolysis temperature increased, energy demand was high and this led to increase in GHG emissions (Alherbawi et al. 2024). Food waste predominantly consisting of cucumber (with high moisture content of 40%) pyrolyzed at 500 °C resulted in very high GHG emissions (ranging from 53,603 to 94,184 ton $CO_{2eq}$/year compared to the other food waste types (Alherbawi et al. 2024).

Using Gabi software to model a pyrolysis process for the conversion of waste date stones into bio-oil, biochar, and syngas, the LCA results showed microwave had 14.94% more impact on climate change, 14.29% more impact on ozone depletion, and 14.44% more photochemical ozone production impact compared to conventional pyrolysis ($-1.74 \times 10^2$ kg $CO_{2eq}$, $-9.80 \times 10^{11}$ kg CFC[14]-11 $_{eq}$, and $-2.77 \times 10^{-1}$ kg

[13] 1,4-Dichlorobenzene per FU (functional unit = 1000 kg).

[14] Chlorofluorocarbon (CFC).

**Table 5.2** Greenhouse gas emissions from Pyrolysis of waste materials

| Type of Waste | Pyrolysis method | Functional unit (FU) | GHG emissions (g $CO_{2eq}$/FU | Reference |
|---|---|---|---|---|
| Cotton straw | Fast pyrolysis for levoglucosan | 1 kg levoglucosan | 4570 | Wang et al. (2020) |
| Lignocellulosic biomass and plastic waste | Microwave pyrolysis into bio-oil, syngas, and biochar | 1 kg bio-oil | 2240–6000 | Foong et al. (2024) |
| Food waste | Microwave pyrolysis into bio-oil | 1 kg bio-oil | 38.92 | Neha et al. (2022) |
| Plastic waste | Microwave pyrolysis for olefins lubricants | 1 kg of olefins lubricants | 1070 | Selvam et al. (2023) |
| Oil palm empty fruit bunches | Fast pyrolysis for bio-oil | 1 kg of bio-oil | 4460 | Chan et al. (2016) |
| Sewage sludge, sawdust, wheat straw | Co-pyrolysis | 1000 kg of bio-oil | 91–680 | Huang et al. (2023) |

NMVOC[15]$_{eq}$ per year for climate change, ozone depletion, and photochemical ozone production respectively) (Parthasarathy et al. 2024). Photochemical ozone arises from products of reactions between OH-radicals, photochemical oxidants, and their air pollutants of nitrogen oxides ($NO_x$) and non-methane volatile organic compounds. This implies that microwave pyrolysis of waste, though technologically efficient, is harmful to the environment compared to conventional pyrolysis due to the high energy requirement. Similarly, Roostaei and Zhang (2017) reported high GHG emissions of $2 \times 10^4$ $kgCO_{2eq}$/50,700 MJ per year for microwave pyrolysis.

Nonetheless, other reports have indicated otherwise. For instance, catalytic microwave pyrolysis of plastic waste contributed about 98.24% GHG emission reduction compared to conventional pyrolysis (Gracida-Alvarez et al. 2023). Table 5.2 shows LCA results of waste pyrolysis.

### *5.4.4 LCA of Waste Torrefaction*

SimaPro software with DATASMART database was used to evaluate the environmental impacts of a mobile forest residues torrefied fuel system for electricity where significant decrease in global warming potential (about 41% GHG emission reduction) was seen when fossil fuels were replaced with torrefied biomass (Alanya-Rosenbaum and Bergman 2019). Again, GHG emissions were reduced by about 8%

[15] Non-methane volatile organic compounds (NMVOC).

when a blend of torrefied biomass (10 wt%) and coal (90%) were used in co-firing plants for electricity generation (Alanya-Rosenbaum and Bergman 2019).

In as much as high energy use in drying units and combustors contributes to GHG emissions, geographic location of the thermochemical conversion facility also has significant environmental burdens as observed by Terry et al. (2024) who carried out LCA study on torrefaction of oil palm empty fruit bunch (EFB) to evaluate GHG emissions on three different siting strategies namely mobile torrefaction system, torrefaction facility closer to palm oil mill, and close proximity of torrefaction system to port. Since transportation contributes significant emissions due to fossil fuel use, it is environmentally beneficial to site torrefaction plants closer to raw material source (Terry et al. 2024).

Forest residues such as acacia, bamboo, and teak were torrefied into charcoal, and using LCA, the environmental impact showed that the use of fertilizer on farmlands for cultivation of these plants had significant burden on human health, ecosystem, resource depletion, and global warming, contributing about 85% of the total environmental impacts considering a cradle-to-gate production of 1 MJ (Partey et al. 2017). An integrated torrefaction system with waste-to-power facility generating about 136,078 Mt/y (metric ton per year) of bio-coal and 28.1 MW heat energy was assessed in terms of environmental sustainability where LCA results showed that GHG emissions reduction of 8.5–25.6% could be seen when about 10–30% of the produced bio-coal is co-fired with fossil-based coal in a power plant for electricity generation, compared to 100% coal-fired electricity generation (Kaliyan et al. 2014). In this torrefaction facility, GHG emissions for electricity generated from 100% bio-coal (to be used by the facility) is about 135.4 g $CO_{2eq}$/kWh which is about 28% lower than when electricity is purchased from the grid (fossil-based). Comparing these results to that of a torrefaction facility where there is no integrated bio-coal-to-energy facility, about 229.5 g $CO_{2eq}$/kWh GHG emissions are released into the atmosphere implying that the integrated system has better environmental performance compared to only torrefaction system (Kaliyan et al. 2014).

## 5.5 Improvement Options for Thermochemical Conversion Technologies

Minimizing electricity utilization (especially fossil-based) in thermochemical conversion facilities would improve the exergetic performance of the systems. The main contributors of exergy destruction in thermal treatment facilities include equipment and lack of recycling and carbon capture. Mechanical equipment like compressors and pumps contribute about 2–4% of the total exergy destroyed in the plant (Peters et al. 2014). Reduction in dryer and milling energy consumptions through equipment improvement such as increasing the screen size of the mill though it may have few challenges with reactors which cannot operate efficiently with large particle size of waste materials would help improve the thermodynamic sustainability of thermochemical conversion technologies. On the other hand, drying to reduce moisture content of waste materials would make them more brittle and drastically reduce energy requirements by the grinder or mill (Mani 2005) but increase the exergy destruction in the dryer.

Exergy destruction in thermal treatment combustors results from chemical reactions of the waste materials. Thus, in order to improve the exergy performance of combustors, the elevation of inlet stream temperature via preheating is recommended though it may generate some exergy destructions for the intermediate unit operation.

Dryers in thermochemical waste conversion facilities can be improved in terms of energy and exergy by reducing the mass flow rate which would drastically reduce exergy destruction caused by pressure drop. Nonetheless, exhaust gas temperature may drastically fall due to the slow charge of drying gas into the dryer (Peters et al. 2014).

## 5.6 Conclusion

Exergy and life cycle assessments have been efficient sustainability analysis tools to pinpoint locations of imperfection in a production system as basis for improvement. Thermochemical waste conversion technologies are found to be exergetically efficient compared to most fossil-based conventional energy generation technologies as discussed in this chapter. The exergetic life cycle assessment of commonly employed thermochemical conversion technologies as well as recent advanced technologies were reviewed, and some improvement options were suggested. Greenhouse gas emissions (GHG) resulting in climate change were the most assessed environmental impact category due to the use of fossil-based fuels in almost all the conventional thermal conversion technologies except for renewable energy hybrid systems.

## References

Afzal S, Singh A, Nicholson SR, Uekert T, DesVeaux JS, Tan ECD et al (2023) Techno-economic analysis and life cycle assessment of mixed plastic waste gasification for production of methanol and hydrogen. Green Chem 25(13):5068–5085

Aghapour sabagh F, Hossainpour S, Pourhemmati S (2023) Energy, exergy, and economic evaluation of integrated waste incineration facility with a thermal power plant. Energy Convers Manag 20:100434

Aich W, Hammoodi KA, Mostafa L, Saraswat M et al (2024) Techno-economic and life cycle analysis of two different hydrogen production processes from excavated waste under plasma gasification. Process Saf Environ Prot 184:1158–1176

Alanya-Rosenbaum S, Bergman RD (2019) Life-cycle impact and exergy based resource use assessment of torrefied and non-torrefied briquette use for heat and electricity generation. J Clean Prod 233:918–931

Alherbawi M, Parthasarathy P, Elkhalifa S, Al-Ansari T, Gordon M (2024) Techno-economic and environmental analyses of the pyrolysis of food waste to produce bio-products. Heliyon 10(6):e27713

AlNouss A-ATA, McKay G (2018) Optimum utilization of biomass for the production of power and fuels using gasification. Comp Aided Chem Eng 43:1481–1486

Andreasi Bassi S, Biganzoli F, Ferrara N, Amadei A, Valente A, Sala SAF (2023) Updated characterisation and normalisation factors for the Environmental Footprint 3. 1 method. Publications Office of the European Union, Luxembourg. https://doi.org/10.2760/798894

Arena U, Parrillo F, Ardolino F (2023) An LCA answer to the mixed plastics waste dilemma: Energy recovery or chemical recycling? Waste Manag 171:662–675

Ayres RU, Ayres LW, Martinás K (1998) Exergy, waste accounting, and life-cycle analysis. Energy 23(5):355–363

Ayres RU, Turton H, Casten T (2007) Energy efficiency, sustainability and economic growth. Energy 32:634–648

Ayub Y, Zhou J, Ren J, Wang Y et al (2023) Plasma gasification based monetization of poultry litter: system optimization and comprehensive 5E (Energy, Exergy, Emergy, Economic, and Environmental) analysis. Energy Convers Manag 282:116878

Bhattacharya A, Das A, Datta A (2014) Exergy based performance analysis of hydrogen production from rice straw using oxygen blown gasification. Energy 69:525–533

Chan YH, Tan RR, Yusup S, Lam HL, Quitain AT (2016) Comparative life cycle assessment (LCA) of bio-oil production from fast pyrolysis and hydrothermal liquefaction of oil palm empty fruit bunch (EFB). Clean Technol Environ Policy 18(6):1759–1768

Chen H, Zhao L, Cong H, Li X (2022) Synthesis of waste heat recovery using solar organic Rankine cycle in the separation of benzene/toluene/p-xylene process. Energy 255:124443

Cornelisse R (1997) The method of exergy analysis. PhD dissertation, University of Twente, The Netherlands

De Meester B, Dewulf J, Janssens A, Van Langenhove H (2006) An improved calculation of the exergy of natural resources for exergetic life cycle assessment (ELCA). Environ Sci Technol 40:6844–6851

Dhaundiyal A, Atsu D (2021) Exergy analysis of a pilot-scale reactor using wood chips. J Clean Prod 279:123511

Dincer I, Zamfirescu C (2014) Integrated power generating systems. In: Zamfirescu C (ed) Dincer I. Advanced Power Generation Systems, Elsevier, pp 455–516

Doherty W, Reynolds A, Kennedy D (2013) Aspen Plus simulation of biomass gasification in a steam blown dual fluidized bed. In: Méndez-Vilas A (ed) Materials and processes for energy: communicating current research and technological developments. Formatex Research Centre, pp 212–220

Foong SY, Chan YH, Yek PNY, Lock SSM et al (2024) Microwave-assisted pyrolysis in biomass and waste valorisation: insights into the life-cycle assessment (LCA) and techno-economic analysis (TEA). Chem Eng J 151942. https://doi.org/10.1016/cej.2024.151942

Gracida-Alvarez UR, Benavides PT, Lee U, Wang M (2023) Life-cycle analysis of recycling of post-use plastic to plastic via pyrolysis. J Clean Prod 425:138867

Granados DA, Velásquez HI, Chejne F (2014) Energetic and exergetic evaluation of residual biomass in a torrefaction process. Energy 74:181–189

Gutierrez MJF, Baxter D, Hunter C, Svoboda K (2005) Nitrous oxide (N2O) emissions from waste and biomass to energy plants. Waste Manag Res 23:133–147

Han X, Chen H, Zhou W, Liang B, Pang S, Du B, Zeng L (2024) Occurrence, distribution and annual emissions of chlorinated paraffins in hazardous byproducts from municipal solid waste incineration plants in South China. Sci Total Environ 925:171764

Hasan MM, Rasul MG, Jahirul MI, Mofijur M (2024) Fuelling the future: unleashing energy and exergy efficiency from municipal green waste pyrolysis. Fuel 357(Part A):129815

Hosseinzadeh-Bandbafha H, Fallahi A, Ghasemkhani H, Shafiei M et al (2022) Exergetic sustainability evaluation of horse manure biomass valorization by microwave pyrolysis. Fuel 323:124286

Huang C, Mohamed BA, Li LY (2023) Comparative life-cycle energy and environmental analysis of sewage sludge and biomass co-pyrolysis for biofuel and biochar production. Chem Eng J 457:141284

International Energy Agency (IEA) (2023) Achieving the COP28 target of tripling global renewable capacity by 2030 hinges on policy implementation. https://www.iea.org/reports/renewables-2023/executive-summary. Accessed 11 Mar 2024

Iribarren D, Susmozas A, Petrakopoulou F, Dufour J (2014) Environmental and exergetic evaluation of hydrogen production via lignocellulosic biomass gasification. J Clean Prod 69:165–175

Jiao L, Li J, Yan B, Chen G, Ahmed S (2022) Microwave torrefaction integrated with gasification: energy and exergy analyses based on Aspen Plus modeling. Appl Energy 319:119255

Kaliyan N, Morey RV, Tiffany DG, Lee WF (2014) Life cycle assessment of a corn stover torrefaction plant integrated with a corn ethanol plant and a coal fired power plant. Biomass Bioenergy 63:92–100

Kotas TJ (1985) The exergy method of thermal plant analysis. Butterworths, London

Lee B, Im S-K (2024) Energy, exergy, and exergoeconomic analyses of plastic waste-to-energy integrated gasification combined cycles with and without heat recovery at a gasifier. Appl Energy 355:122243

Lee KT, Ofori-Boateng C (2013) Sustainability of biofuel production from oil palm biomass. Springer Science+Business Media, Singapore, pp 323

Li R, Song G, Huang D, Hu S, Fantozzi F, Hassan A et al (2024) Comparative study of process simulation, energy and exergy analyses of solar enhanced char-cycling biomass pyrolysis process. Energy Convers Manag 302:118082

Li X, Xu H, Gao Y, Tao Y (2010) Comparison of end-of-life tire treatment technologies: a Chinese case study. Waste Manag 30:2235–2246

Lin PQ, Yu X, Wang H, Ming H, Ge S, Liu F, Peng H, Sonne C, Zhang L (2023) Life cycle assessment of bio-oil prepared from low-temperature hydrothermal oxide-catalyzed cotton stalk. Energy. https://doi.org/10.1016/j.energy.2023.128554

Manatura K, Lu J-H, Wu K-T, Hsu H-T (2017) Exergy analysis on torrefied rice husk pellet in fluidized bed gasification. Appl Therm Eng 111:1016–1024

Mani S (2005) A systems analysis of biomass densification process. University of British Columbia. https://doi.org/10.14288/1.0058759

Mohamed BA, Ruan R, Bilal M, Periyasamy S, Awasthi MK, Rajamohan N, Leng L (2024) Sewage sludge co-pyrolysis with agricultural/forest residues: a comparative life-cycle assessment. Renew Sustain Energy Rev 192:114168

Mulya KS, Zhou J, Phuang ZX, Laner D, Woon KS (2022) A systematic review of life cycle assessment of solid waste management: methodological trends and prospects. Sci Total Environ 831:154903

Neha S, Ramesh KPK, Remya N (2022) Techno-economic analysis and life cycle assessment of microwave co-pyrolysis of food waste and low-density polyethylene. Sustain Energy Technol Assess 52(Part D):102356

Neri E, Berti B, Passarini F, Vassura I, Giorgini L, Zattini G, Tosi C, Cavazzoni M (2018) Application of LCA methodology in the assessment of a pyrolysis process for tyres recycling. Environ Eng Manag J 17:2437–2445

Nivash V, Sakthivadivel D, Alaswad A, Vigneshwaran V S (2024) Techno-economic, energy, and exergy analyses of invasive weed gasification for hydrogen enriched producer gas production. Heliyon 10(6):e27673

Ochoa-Garcia M, Tejeda-Lopez L, Ojeda-Delgado K, Gonzalez-Delgado AD, Sanchez-Tuiran E (2018) Process simulation and exergy analysis of microalgal biodiesel production using *Chlorella vulgaris* via $ZnCl_2$ pretreatment. Indian J Sci Technol 11(23):1–12

Ofori-Boateng C (2022) Algal biorefining: resource expenditure and exergo-environmental sustainability. Nova Science Publishers, New York, p 296

Ofori-Boateng C, Lee KT (2014) An oil palm-based biorefinery concept for cellulosic ethanol and phytochemicals production: sustainability evaluation using exergetic life cycle assessment. Appl Therm Eng 62(1):90–104

Ofori-Boateng C, Lee KT, JitKang L (2012) Feasibility study of microalgal and jatropha biodiesel production plants: Exergy analysis approach. Appl Therm Eng 36(1):141–151

Okolie JA, Nanda S, Dalai AK, Kozinski JA (2020) Hydrothermal gasification of soybean straw and flax straw for hydrogen-rich syngas production: experimental and thermodynamic modeling. Energy Convers Manag 208:112545

Ozturk MU, Ayol A, Tezer O (2023) Life cycle assessment of olive pomace gasification for an up-draft fixed bed gasifier system. Int J Hydrog Energy 48(60):23339–23347

Partey ST, Frith OB, Kwaku MY, Sarfo DA (2017) Comparative life cycle analysis of producing charcoal from bamboo, teak, and acacia species in Ghana. Int J Life Cycle Assess 22(5):758–766

Parthasarathy P, Tahir F, Pradhan S, Al-Ansari T, McKay G (2024) Life cycle assessment of biofuel production from waste date stones using conventional and microwave pyrolysis. Energy Convers Manage X(21):100510

Peralta-Ruiz Y, Obregon LG, González-Delgado Á (2018) Design of biodiesel and bioethanol production process from microalgae biomass using exergy analysis methodology. Chem Eng Transact 70:1045–1050

Peters JF, Petrakopoulou F, Dufour J (2014) Exergetic analysis of a fast pyrolysis process for bio-oil production. Fuel Process Technol 119:245–255

Peters JF, Banks SW, Bridgwater AV, Dufour J (2017). A kinetic reaction model for biomass pyrolysis processes in Aspen Plus. Appl Energy 188:595–603

Prakash P, Furqan T, Snigdhendubala P, Tareq A, McKay G (2024) Life cycle assessment of biofuel production from waste date stones using conventional and microwave pyrolysis. Energy Convers Manag: X 21:100510

Rispoli AL, Verdone N, Vilardi G (2021) Green fuel production by coupling plastic waste oxy-combustion and PtG technologies: economic, energy, exergy and CO2-cycle analysis. Fuel Process Technol 221:106922

Rivero R, Rendon C, Monroy L (1999) The exergy of crude oil mixtures and petroleum fractions: calculation and application. Int J Thermodyn 2(3):115–123

Roostaei J, Zhang Y (2017) Spatially explicit life cycle assessment: opportunities and challenges of wastewater-based algal biofuels in the United States. Algal Res 24:395–402

Rosen MA, Dincer I (2003) Exergy-cost-energy-mass analysis of thermal systems and processes. Energy Convers Mgmt 44:1633–1651

Sanjaya E, Abbas A (2023) Plasma gasification as an alternative energy-from-waste (EFW) technology for the circular economy: an environmental review. Resourc Conserv Recycl 189:106730

Selvam E, Kots PA, Hernandez B, Malhotra A, Chen W, Catala-Civera JM, Santamaria J, Ierapetritou M, Vlachos DG (2023) Plastic waste upgrade to olefins via mild slurry microwave pyrolysis over solid acids. Chem Eng J 454:140332

Shi T, Moktadir MdA, Ren J, Shen W (2023) Comparative economic, environmental and exergy analysis of power generation technologies from the waste sludge treatment. Energy Convers Manag 286:117074

Singh RK, Jena K, Chakraborty JP, Sarkar A (2020) Energy and exergy analysis for torrefaction of pigeon pea stalk (cajanus cajan) and eucalyptus (*eucalyptus tereticornis*). Int J Hydrog Energy 45(38):18922–18936

Skabelund BB, Jenkins CD, Stechel EB, Milcarek RJ (2023) Thermodynamic and emission analysis of a hydrogen/methane fueled gas turbine. Energy Convers Manag: X 19:100394

Szargut J (1988) Energy and exergy analysis of the preheating of combustion reactants. Int J Energy Res 12:45–58

Szargut J (2005) Exergy method. WIT Press, Southampton, UK, Technical and Ecological Applications
Tang Y, Dong J, Li G, Zheng Y, Chi Y, Nzihou A, Weiss-Hortala E, Ye C (2020) Environmental and exergetic life cycle assessment of incineration- and gasification-based waste to energy systems in China. Energy 205:118002
Teles dos Santos M, Park SW (2009) Exergy and sustainable development for chemical industry revisited. Comput Aided Chem Eng 27:1923–1928
Temireyeva A, Sarbassov Y, Shah D (2024) Process simulation of flax straw pyrolysis with kinetic reaction Model: experimental validation and exergy analysis. Fuel 367:131494
Terry LM, Chew JJ, How BS, Sunarso J, Aqsha A (2024) Life cycle assessment of empty fruit bunch torrefaction on greenhouse gases emission. In: Abraham MA (ed) Encyclopedia of sustainable technologies (second edition). Elsevier, pp 302–322
Trindade AB, Palacio JCE, González AM, Orozco DJR et al (2018) Advanced exergy analysis and environmental assesment of the steam cycle of an incineration system of municipal solid waste with energy recovery. Energy Convers Manag 157:195–214
Tsangas M, Papamichael I, Loizia P, Voukkali I, Raza NS, Vincenzo N, Zorpas AA (2024) Life cycle assessment of electricity generation by tire pyrolysis oil. Process Saf Environ Prot 186:376–387
Tsangas M, Papamichael I, Banti D, Samaras P, Zorpas AA (2023) LCA of municipal wastewater treatment. Chemosphere 341:139952
UN Environment Program (2019) Emissions Gap Report 2019 ISBN: 978–92–807–3766–0 Job number: DEW/2263/NA. https://wedocs.unep.org/bitstream/handle/20.500.11822/30797/EGR2019.pdf?sequence=1&isAllowed=y. Accessed 4 Jul 2023
Valero A, Lozano MA, Muñoz M (1986) A general theory of exergy saving. ASME book No. H0341A, WAM AES. New York: ASME, pp 1–22
Vilardi G, Verdone N (2022) Exergy analysis of municipal solid waste incineration processes: the use of O2-enriched air and the oxy-combustion process. Energy 239(Part B):122147
Wang J, You S, Lu Z, Chen R, Xu F (2020) Life cycle assessment of bio-based levoglucosan production from cotton straw through fast pyrolysis. Bioresour Technol 307:123179
Wang L, Teles PR, Haoshui Y, Brenda VFS, Ahmad A (2024) Waste incineration and heat recovery hybridized with low-focus fresnel lens solar collectors for sustainable multi-generation; a thorough techno-economic-environmental analysis and optimization. Chemosphere 346:140409
Xiang X, Gong G, Wang C, Cai N, Zhou X, Li Y (2021) Exergy analysis of updraft and downdraft fixed bed gasification of village-level solid waste. Int J Hydrog Energy 46(1):221–233
Xiao Z, Wu W (2024) Environmental sustainability by an integrated system based on municipal solid waste fluidized bed gasification using energy, exergy, and environmental analyses. Appl Therm Eng 239:122171
Yan W, Hastings JT, Acharjee TC, Coronella CJ, Vásquez VR (2010) Mass and energy balances of wet torrefaction of lignocellulosic biomass. Energy Fuels 24:4738–4742
Yang H, Guo Y, Fang N, Dong B (2023) Life cycle assessment of greenhouse gas emissions of typical sewage sludge incineration treatment route based on two case studies in China. Environ Res 231(Part 1):115959
Yang S, Liang A, Peng S, Liu Z, Deng C, Xie N (2024) Energy, exergy and exergo-economic analysis of a novel HT-PEMFC based cogeneration system integrated with plasma gasification of food waste. Appl Therm Eng 246:122924
Zhao C, Huang J, Xie D, Qiao Y, Xu M (2024) Thermodynamic and techno-economic analysis of hydrogen production from food waste by torrefaction integrated with steam gasification. Energy Convers Manag 299:117826
Zhou C, Hu D, Wang R, Liu J (2011) Exergetic assessment of municipal solid waste management system in south Beijing. Ecol Complex 8(2):171–176

Zhuo C, Xueqin L, Zhiwei W, Yantao Y, Tanglei S et al (2023) Techno-economic and whole life cycle assessment of ester fuels production from agricultural waste via hydrothermal liquefaction. Ind Crops Prod 192:116096

Zoppi G, Tito E, Bianco I, Pipitone G, Pirone R, Bensaid S (2023) Life cycle assessment of the biofuel production from lignocellulosic biomass in a hydrothermal liquefaction—aqueous phase reforming integrated biorefinery. Renewable Energy 206:375–385

# Chapter 6
# Technoeconomic Analysis of Thermochemical Waste Conversion Technologies

**Abstract** Thermochemical waste conversion technologies like pyrolysis and gasification have gained much attention over the years due to their high efficiencies and the capability to handle diverse waste composition in co-production processes. One of the effective ways to transform waste materials (e.g. plastic waste, waste tires, agricultural waste, etc.) into value-added bioproducts like bio-oil, bioelectricity, and biochemicals is through waste-to-energy (WtE). The technoeconomic analysis (TEA) results of thermochemical waste conversion technologies were discussed and compared with fossil-based conversion technologies. Bioproducts from thermochemical conversion technologies are found to be highly competitive with fossil diesel in terms of cost though they require high initial capital investment (TCI) for commercial scale facilities. The scale-up of waste pyrolysis, for instance, is found to drastically reduce the cost of bio-oil by about 35–88% which could be competitive with fossil-based diesel. Sensitivity analysis results of thermochemical waste conversion technologies are discussed in this chapter.

**Keywords** Sustainability · Thermochemical conversion technology · Technoeconomic analysis · Pyrolysis · Gasification · Waste management

## 6.1 Introduction

Waste management usually involves technologies which must be sustainable environmentally, economically, and technologically, not neglecting socio-political sustainability for a unified clean and sound community. Biofuel production currently is sourced from first-generation feedstock (i.e., edible crops like corn, canola oil, etc.) which competes with global food production and security. In order to break the food versus fuel delineation as well as mitigate waste disposal in the environment, second-generation feedstock (which forms part of waste materials) has evolved. As a result, waste materials such as agricultural and forest residues, plastic and tire wastes, municipal solid waste (MSW), industrial and medical waste, municipal liquid waste



C. Ofori-Boateng, *Sustainability of Thermochemical Waste Conversion Technologies*,
https://doi.org/10.1007/978-3-031-64342-2_6

(MLW), etc. have become cost-efficient and environmentally sustainable feedstock for biofuel production.

Recent trends in thermochemical conversion technologies focus mainly on process optimization, technological advancement, and environmental sustainability. However, there are only a few reports in literature that address economic sustainability of biofuel production technologies. For a successful process scale-up for waste pyrolysis, torrefaction, gasification, hydrothermal treatment, combustion, and incineration, it is imperative to assess the technoeconomic performance to ascertain the cost benefit and feasibility of the technology.

Technical assessment coupled with economic analysis of a system can be a unique driving force for continuous process improvement which can be used to forecast the overall cost of production via experimental laboratory and pilot scale-up methods (You et al. 2008). Today's focus of economic analysis of biofuel technologies dwells on increasing man-made capital stocks as opposed to the traditional method of determining the cost of a process and a product which aims to efficiently allocate resources and equitable distribution (OECD 2001).

Thermochemical conversion technology is considered economically viable when it can accommodate the existing conditions of a society which supports it by protecting the natural environment and financial resources available. Economic sustainability of a technology depends on the quantity of natural resources it utilizes to generate goods and provide services that would benefit society and would not jeopardize the environment with its activities. A thermal conversion technology that helps to manage waste like MSW by utilizing them for value-added bioproducts may not be fully considered economically sustainable if the emissions generated are unaccounted for as well as the products formed are costly. Hence, economic sustainability of a thermochemical waste conversion technology can be attained when tactical options of connecting the linkages between eco-friendly environment, cost effective goods and services as well as judicious use of natural resources for a peaceful society today and the future are considered.

The aim of this chapter is to distinctly spell out the cost benefits and efficiencies of thermochemical waste conversion technologies and compare economic sustainability results for the different technologies.

## 6.2 Thermochemical Conversion Technologies: Introduction to Technoeconomic Analysis

Technoeconomic Analysis (TEA) is a commonly used economic sustainability assessment tool to evaluate processes, products, and systems to know the major economic inference of the prospects of transforming raw materials into value-added products (Kannah et al. 2021). TEA is integrated into thermochemical conversion technologies to aid in the selection of appropriate methods for the successful development, operation, maintenance, and policy making decisions, since TEA results may inform a community about the influence of a product or technology on job creation, energy security, economic stability, inflation, and tax implications (Farzad et al. 2021).

TEA can be carried out using Microsoft programs, spreadsheet software (e.g. excel), or economic analysis process simulators or tools like Life Cycle Cost (LCC), four-level economic potential approach, PYROL HYSYS, Pro/II simulator, CHAM-CAD, ASPEN Hysys (ASPEN Plus), SuperPro Software, Monte Carlo simulation model, Financial and Risk analysis simulator, Python-based BioSTEAM, etc. (Meyer et al. 2020; Do and Lim 2016; Michailos et al. 2017). TEA has been used to evaluate the cost benefits of thermochemical waste conversion processes such as pyrolysis, gasification, etc. for the production of bio-oil, syngas, and biochar using economic indicators that are based on important economic factors (Economopoulos 2010; Yousef et al. 2023; Farzad et al. 2021). The sustainability analysis of thermochemical conversion technologies can be performed using different kinds of sustainability models such as triple-bottom-line analysis model,[1] pay-as-you-go approach[2] etc. (Sharma et al. 2019). The main economic factors used in TEA evaluation of biofuel and bioenergy system are discussed in the next subsections.

### *6.2.1 Operating Cost*

Operating Cost (OC) is the cost incurred during the operation of the plant/facility. For a thermochemical conversion facility, the operating cost may include the cost of raw material (e.g., waste, gasifying agent, chemicals, etc.), energy, factory expenses, administrative, disposal, and labor. To keep operating cost stable and at a minimum, there must be consistent monitoring and optimization of energy utilization as well as waste material availability in thermochemical conversion facilities since these technologies such as gasification, pyrolysis, etc. are known to be energy-intensive. With a known flow rate or operating capacity of the plant, the total operating cost can be estimated using Eq. 6.1.

$$\mathrm{TOC} = \mathrm{OC} + \sum_{i=1}^{n} \frac{\mathrm{OR} \times \mathrm{PC}}{(1+r)^{i}} \tag{6.1}$$

where

TOC = total operating cost,
OC = operating cost
OR = operating rate

[1] The triple bottom line is a sustainability framework which measures a plant's operational success based on profit, people, and the planet.

[2] Pay-as-you-go approach is a sustainability framework which uses resource optimization to align costs with actual usage, reduce wastage, and promote environmentally sustainable cost-efficient products and services.

PC = annual production capacity
$r$ = interest rate, $i$ = number of years, $n$ = plants lifetime.

### 6.2.2 Capital Cost (CC)

Capital Cost (CC) is the one-time or fixed upfront expenditure, or investment spent to finance a business, and it consists of the cost of acquiring equipment, land, instrumentation, contingencies and fees, construction, installations, and other assets acquisition costs. The capital cost depends largely on the production capacity of the plant. Basically, with rise in production capacities of thermochemical conversion technologies, the capital cost per unit of production also decreases. This implies that, commercial scale facilities may have high economies of scale compared to experimental laboratory scale, due to purchase discounts for equipment and supplies, high automation, and efficiency hence lower cost of labor per unit of production. However, small-scale thermal conversion facilities may have low capital cost and can have cost benefits depending on other factors. Nonetheless, considering the total capital cost of biofuel production technologies without considering the cost per unit production, commercial large-scale plants have high initial capital cost compared to small-scale plants (Khan et al. 2023). For example, a laboratory scale pyrolysis of waste requires an estimated capital cost of about 1000–15,000 depending on the capacity compared to capital cost of about US $500,000 to US $5 million for capacities up to 70,000 metric tons (Khan et al. 2023; Kodera et al. 2021).

The ASPEN Plus model for economic assessment of thermochemical conversion technologies can be used to estimate the components of the major factors that affect the economic feasibility of thermal facilities. The main economic sustainability evaluation indicators are the total capital investment (TCI), total production cost (TPC), sensitivity and profitability indicators. TCI encompasses the capital cost (CC), and it consists of the ***fixed capital investment*** (FCI) and ***Working Capital*** (WC).

The FCI, also called the ***Bare Module Cost*** ($C_{BM}$), is the cost of investment needed for the supply of construction overheads and plant components that are directly or indirectly related to the production (Seider et al. 2008). WC is the amount of money needed to start the project which is usually estimated as 0.15 times the FCI according to Sinnot (1986) and Peters and Timmerhaus (1981). FCI of pyrolysis and gasification plants were estimated at €11.18 million and €14.44 million respectively, and are found to be higher compared to incineration facility FCI (€22.15 million) (Danish Energy Agency 2023; Fadhilah et al. 2023).

TPC is the total cost needed to operate the facility. This is basically the operating cost, including the marketing of the products. Feedstock and other raw material cost forms about 40–45% of the TPC of a biomass/waste-to-energy facility which is lower compared to that of coal-to-energy facilities (50–52% TPC is raw material cost) (Wang et al. 2019). According to Jain and Sharma (2010), TPC consists of:

- Fixed costs, which include maintenance, supervision, insurance, and royalties.

- Variable cost, which includes cost of raw materials, utilities, shipping, etc. For thermochemical conversion technologies, variable cost can include the costs of gasifying agents, waste materials, energy requirement (Fadhilah et al. 2023).
- General expenses, which consist of administrative, engineering, and legal costs, distribution, and sales, etc.

Equation 6.2 is used for the estimation of equipment cost (Seider et al. 2008).

$$C_{\mathrm{BM}} = C_p^0 F_{\mathrm{BM}} \tag{6.2}$$

where

$C_{\mathrm{BM}}$ = Bare Module Cost
$F_{\mathrm{BM}}$ = Bare Module Factor, which is estimated using Eq. 6.3
$C_p^0$ = purchase cost of base conditions, which is estimated using Eq. 6.5

$$F_{\mathrm{BM}} = B_1 + B_2 F_M F_P \tag{6.3}$$

where

$B_1$ and $B_2$ are constants which depend on the type of equipment
$F_M$ = material factor which depends on the equipment type.
$F_P$ = pressure factor which is defined by Eq. 6.4:

$$\log_{10} F_P = C_1 + C_2 \log_{10} P + C_3 \left(\log_{10} P\right)^2 \tag{6.4}$$

where

$C_1$ and $C_2$ are constants depending on the type of equipment and material it is made of.

$$\log_{10} C_p^0 = K_1 + K_2 \log_{10}(A) + K_3\left[\log_{10}(A)\right]^2 \tag{6.5}$$

where

$C_p^0$ = purchase cost
$K_1$, $K_2$, and $K_3$ are constants depending on the equipment type.
$A$ = capacity factor or the size of the equipment.

Parameters like $K_1$, $B_{1,}$ and $C_1$ which are not found in the database of ASPEN Plus have been provided elsewhere (Lim et al. 2009; Turton et al. 2009).

The life cycle cost (LCC) assessment model has indicators similar to ASPEN Plus model. Equation 6.6 summarizes the major components of LCC analysis for biofuel technologies (Lee and Ofori-Boateng 2013):

$$\mathrm{LCC} = \mathrm{Capital\ Cost(CC)} + \mathrm{Operating\ Cost(OC)}$$

$$+ \text{Maintenance Cost(MC)} + \text{Feedstock Cost(FC)}$$
$$- \text{Salvage Value(SV)} - \text{By - product credits(BPC)} \tag{6.6}$$

In most circumstances, the ***feedstock cost*** (FC) or ***feedstock consumption*** (FC) has a drastic impact on the cost of products. For thermochemical conversion of waste, the feedstock (i.e., the waste materials) may be obtained at zero cost hence reducing the final cost of bio-oil, biochar, or syngas compared to using first-generation feedstocks like corn or third-generation feedstock like algae biomass. However, there are many instances where the procurement of waste may be associated with a gate fee which varies depending on location, waste composition, etc. (Hogg 2002). Paida et al. (2019) suggested a constant value of US$ 75 per ton of organic waste as the cost of feedstock in TEA evaluation of hydrothermal gasification system for hydrogen production. The cost of feedstock consumed in the facility over its lifespan can be estimated using Eq. 6.7 knowing the price of waste and the rate of waste consumption.

$$\text{FC} = \sum_{i=1}^{n} \frac{\text{FP} \times \text{FU}}{(1+r)^i} \tag{6.7}$$

where

FC = feedstock consumption
FP = feedstock price
FU = feedstock cost
$r$ = interest rate, $i$ = number of years, $n$ = plants lifetime.

***Salvage value*** (SV) is the projected estimate of the technology or process at the end of its useful life. It is the remaining value of the facilities' components and other assets at the end of the plant's lifespan. SV can be estimated based on current market demand and the value of similar technologies in the market considering additional costs for disposal or recycling. SV can have an impact on the economic sustainability of thermochemical conversion technology which can propose informed decision for investors and stakeholders. Based on depreciation rate a model that considers replacement cost other than initial capital cost (Eq. 6.8) is used to estimate the SV.

Equation 6.9 is also used to project the present value of the SV.

$$\text{SV} = \text{RC}(1-d)^{n-1} \tag{6.8}$$

$$\text{SV}_{\text{PV}} = \frac{\text{RC} \times (1-d)^{n-1}}{(1+r)^n} \tag{6.9}$$

where

SV = *Salvage* value
RC = *replacement* cost
$\text{SV}_{\text{PV}}$ = present value of salvage cost

$r$ = interest rate, $d$ = depreciation ratio, $n$ = plants lifetime.

***By-product credits*** (BPC) are credits (or money) obtained from the recycling, reuse, or sales of by-products such as biochar, bottom ash, tar, etc. generated during thermal treatment of waste. BPC can be used to balance out some of the cost associated with biofuel production and facilities that benefit from BPC contribute tremendously to waste management and emissions reduction. In a pyrolysis facility for instance, the BPC can be estimated (Eq. 6.10) with the assumption that the biochar generated would have a fixed price at the time of estimation, whilst knowing the production capacity and the biochar/syngas conversion factor over the pyrolysis plant's lifespan.

$$\mathrm{BP} = \sum_{i=1}^{n} \frac{\mathrm{GP} \times \mathrm{GCF} \times \mathrm{PC} \times 1000}{(1+r)^i} \tag{6.10}$$

where

BP = by-product (biochar and syngas) credit
GP = by-product price (biochar and syngas)
GCF = by-product (biochar and syngas) conversion factor from waste material
PC = annual pyrolysis oil production capacity
$r$ = interest rate, $i$ = number of years, $n$ = plants lifetime.

In terms of interest rate, Eq. 6.6 can be rewritten as Eq. 6.11 (Lee and Ofori-Boateng 2013).

$$\mathrm{LCC} = \mathrm{CC} + \sum_{i=1}^{n} \frac{\mathrm{OC}_i + \mathrm{MC}_i + \mathrm{FC}_i}{(1+r)^i} - \frac{\mathrm{SV}}{(1+r)^n} - \sum_{i=1}^{n} \frac{\mathrm{BPC}_i}{(1+r)^i} \tag{6.11}$$

where

LCC = Life Cycle Cost
$n$ = lifetime of facility
CC = capital cost
OC = Operating cost
MC = Maintenance cost
FC = Feedstock cost
SV = Salvage value
BPC = By-product credits
$r$ = interest rate, $i$ = number of years

On the other hand, LCC may be estimated from Eqs. 6.12 and 6.13

$$\mathrm{LCC} = C_i + P_{V_r} + P_{V_s} \tag{6.12}$$

$$\text{LCC} = C_i + \left[\sum_{1}^{n} C_{ap}.\left[\frac{x(1-x)}{1-x}\right]\right] + \left[C_s.\left[\frac{1+i}{1+d}\right]^n\right] \tag{6.13}$$

where

$C_i$ = initial investment cost which can be the capital cost (CC)
$P_{Vr}$ = recurring annual payment, which involves the constant payment made yearly toward the operation and maintenance of the plant over its lifespan. $P_{Vr}$ may comprise expenses for employee salaries, etc. which may affect the economic sustainability of biofuel and bioenergy plants.
$P_{Vs}$ = single future payment, which is the payment made in the future toward some capital costs as a lump sum over the lifespan of the facility.
$C_s$ = single future cost, which is the present value of all future cost pertaining to the facility over its lifespan. This cost includes energy cost (EC), feedstock cost (FC), operating cost (OC), maintenance cost (MC), etc.
$x = (1 + i)/(1 + d)$, $i$ = interest rate, $d$ = discounted rate, $n$ = number of years for payment.

The interpretation of LCC results can be presented based on cost per annum using annualized life cycle cost (ALCC) or levelized cost (LC) where the LCC results are expressed in terms of total production or unit energy cost over the lifespan of the plant (see Eq. 6.14):

$$\text{Levelized cost} = \frac{\text{LCC}}{\text{TPP}} \tag{6.14}$$

where

LCC = Life cycle cost
TPP = Total Product Production Cost which is the production cost of bio-oil, biochar, syngas, bioelectricity, or biochemicals generated via thermochemical conversion processes like pyrolysis.

To estimate the *present value* (which can be represented by *present worth factor PWF*) of thermochemical conversion technologies, the future cash flow information about the plant must be known which can be used to evaluate how feasible the technology will be based on the current interest rate (see Eq. 6.15). The PWF can be compounded over the lifetime of the plant, and this is presented in Eq. 6.16.

$$\text{PWF} = \frac{1}{(1+r)^i} \tag{6.15}$$

where

PWF = Present worth factor
$r$ = interest rate
$i$ = number of years

$$\mathrm{CPW} = \sum_{i=1}^{n} \frac{1}{(1+r)^i} = \frac{(1+r)^n - 1}{r(1+r)^n} \quad (6.16)$$

where

CPW = Compound present worth,
$r$ = interest rate,
$i$ = number of years,
$n$ = plants lifetime.

### 6.2.3 Maintenance Cost (MC)

The maintenance cost (MC) of a plant is the cost incurred by the facility to keep its assets in good condition. Some examples of maintenance cost associated with thermochemical conversion technologies include the cost of control and instrumentation systems repair, equipment maintenance and repair (for reactors, waste preparation and handling units, furnace/boilers, etc.), and other renovations. The cost of inspection, cleaning, replacement of equipment parts and components, to mention but a few, are part of maintenance costs. Maintenance in a thermal conversion facility is necessary to maintain product efficiency, cost effective products, optimal performance, safety, and reliability of the technology to increase the lifespan of the facility. MC of thermochemical conversion facilities can be estimated using Eq. 6.17.

$$\mathrm{MC} = \sum_{i=1}^{n} \frac{\mathrm{MR} \times \mathrm{CC}}{(1+r)^i} \quad (6.17)$$

where

MC = maintenance cost
MR = maintenance ratio
CC = capital cost
$r$ = interest rate, $n$ = project or plant life, and $i$ = number of year.

### 6.2.4 Profitability and Sensitivity Analysis

Profitability of a thermochemical conversion facility can be estimated using the ***Return on Investment*** (ROI), ***Payback Period*** (PP), ***Discounted cash flow rate of return*** (DCCFRR), ***Net Present or Future Worth/Value*** (NPV, or NPW), ***Breakeven Point*** (BEP), etc. Because thermochemical conversion technologies are energy-intensive and require high initial investment, estimating payback period (PP) of the facility would help know the potential risks and the return on investment (ROI) to determine the economic sustainability of the plant. Payback time or period (PP) is the minimum time it takes for a plant or project's investment to pay for itself through the profits, revenues, and savings accrued. If PP is short, it implies that a faster ROI is forecast with high economic benefit of the technology. Equations 6.18 and 6.19 are used to find the payback time of a production plant based on the LCC and ASPEN Plus models respectively.

$$\mathrm{PP} = \frac{\mathrm{CC}}{\mathrm{TBS} - \mathrm{TPS} - \mathrm{TAX}} \tag{6.18}$$

where

PP = payback period
TBS = annual total biofuel sales
TPC = annual total production cost
TAX = annual total tax.

$$\text{Payback Period} = \frac{(\text{Depreciable Fixed Capital Investment})}{(\text{Average Pofit}) + (\text{Average Depreciation})} \tag{6.19}$$

ROI is the ratio of the average cumulative cash flow to the total initial investment expressed as a percentage. ROI is mathematically defined by Sinnot (1986) in Eq. 6.20:

$$\mathrm{ROI} = \frac{(\text{Cummulative net cash flow at end of plant life})}{(\text{Plant life}) \times (\text{Initial Investment})} \times 100\% \tag{6.20}$$

The NPW or NPV is a financial indicator which considers investment costs, revenues, and operating costs generated by a plant to estimate the value of the production technology in terms of future cash flows. According to Tomomatsu and Swallow (2007), the net cash flow in each year of the plant's life is usually linked to its present worth at the beginning of the production by discounting it at some chosen compound interest rate which is selected to reflect the earning power of money. NPW and total NPW can be calculated using Eqs. 6.21 and 6.22 respectively (Sinnot 1986):

$$\text{Net present worth(NPW)} \frac{\text{Estimated net cash flow in year } n(\text{NFW})}{(1+r)^n} \tag{6.21}$$

$$\text{Total NPW of project} = \sum_{n=1}^{n=t} \frac{\mathrm{NFW}}{(1+r)^n} \tag{6.22}$$

where

$r$ = the discount rate (interest rate) in percentage.
NFW = the future worth of the net cash flow in year $n$,
$t$ = the life of the project (years)

***Break-even point (BEP)*** of a production plant is the time of the plant's life where there is neither net profit nor net loss. At this point, the generated revenue is equal to the operational cost. The break-even point determines if the technology or the plant can pay for its costs and start making profits. Based on the value of the CC, OP, and revenues from sales of products (and by-products), Eq. 6.23 can be used to estimate the BEP of the plant (Sinnot 1986):

$$\text{BEP} = \frac{\text{Fixed Costs}}{\text{Total Sales} - \text{Variable Cost}} \times 100 \tag{6.23}$$

Thermochemical conversion technologies have a high probability of attractive returns on investment because of the wide variety of products they can generate at high yields and efficiencies. A ***discounted cash flow rate of return*** (DCFRR) of a thermal conversion plant is the maximum rate at which the technology or plant could pay for itself and still break-even by the end of the project life. DCFRR is used to calculate the present value of future earnings at a specified interest rate. The interest rate from the NPW estimation for which the cumulative net present worth at the end of the project is zero is termed the discounted cash flow rate of return (DCFRR). Mathematically, DCFRR is defined by Eq. 6.24 (Sinnot 1986; Peters and Timmerhaus 1981):

$$\text{DCFRR} = \sum_{n=1}^{n=t} \frac{\text{NFW}}{(1+r')^n} = 0 \tag{6.24}$$

where

$r'$ = the discounted cash flow rate of return in percentage
NFW = the future worth of the net cash flow in year $n$
$t$ = the life of the project (years)

***Sensitivity analysis*** of thermochemical conversion facilities is used to evaluate the changes in input variables and assumptions made in the economic analysis and their impacts on the results of the model employed. Sensitivity analysis reveals the critical factors that influence the economic and environmental performance of the plant. During sensitivity analysis of a pyrolysis plant for instance, factors such as waste composition, process conditions (e.g. temperature, pressure, heating rate, residence time, etc.), equipment costs and efficiencies as well as the products distribution (oil, char, and syngas yields) are varied to assess their effects on important pyrolysis parameters like conversion efficiencies, greenhouse gas (GHG) emissions, energy output. From sensitivity analysis of a pyrolysis plant, scientists, investors, and stakeholders would be able to comprehend the opportunities and issues associated with the implementation of the thermochemical waste conversion plant operation.

## 6.3 Technoeconomic Analysis of Common Thermochemical Waste Conversion Technologies

Hydrogen gas ($H_2$) from waste as discussed in *Chapter 3*, is an eco-friendly biofuel whose utilization or consumption has seen a drastic rise over the years and is reported to contribute about 8% of global energy demand at a production cost of US$ 2.50/kg which is projected to decrease by about 39% by 2030 (Hydrogen Council 2020). Depending on the raw material, the type of thermochemical conversion technology as well as the location of plant, the cost of $H_2$ may differ from region to region. $H_2$ from all sources and generated via different technologies apart from gasification and pyrolysis may sell at US$ 0.8–4/kg at different regions of the world based on specific conditions.

$H_2$ can be produced from biochemical conversion technologies also for cost effective system but depends largely on energy sourced from fossil fuels for its production (Wang and Zhang 2017). Many studies have reported results of TEA of hydrogen production from thermochemical conversion technologies especially gasification (Wang et al. 2019; Gul et al. 2021; Kannah et al. 2021; Badger et al. 2011). Though pyrolysis generates mainly oil, the gases produced as by-products can be upgraded into hydrogen if the process conditions enable high formation of hydrogen in the gas stream/product.

Gasification and pyrolysis involve the use of high efficiency reactors which add significant costs to the economic performance of the plant. Nonetheless, despite their high capital cost, there has been global increasing attention in thermochemical waste-to-energy plants compared to conventional or other conversion technologies' waste-to-energy plants because of their high conversion efficiencies, low environmental impacts, better handling of different components of waste mixture as well as their high economic and exergetic performances (Boie 1953; Ouda et al. 2016).

Gasification of biomass in waste-to-energy facilities is reported to consume high quantities of raw materials (gasifying agents, waste materials, etc.) with high TCI compared to coal-to-energy facilities (Wang et al. 2019). However, the production cost for the biomass gasification system was about 17.3% lower than that of the coal gasification system due to the cheaper cost of feedstock for the biomass gasification system (Wang et al. 2019).

A biomedical waste-to-energy incineration system (pyrolysis and gasification) with production capacities of 27.16 ton/h (79.97 MW) is estimated to consume about 0.011 kWh/kg and 0.096 kWh/kg energy in the waste preparation/pretreatment (drying, pelletizing, shredding, etc.) section of the pyrolysis and gasification units respectively (Gul et al. 2021; Badger et al. 2011). Yousef et al. (2023) compared the large-scale data with small laboratory scale studies of 0.2 kg/h and 1 kg/h capacities for pyrolysis and gasification respectively, and the results showed that about 872.74 kWh energy is required to transform 1 kg of biomedical waste into bioproducts (42.3% pyrolysis oil and 89.7% hydrogen-rich syngas from gasification). At 500 °C pyrolysis (fluidized bed reactor, $N_2$ gas at 20 L/min for 1 h) and 800 °C gasification (bubbling fluidized bed gasifier, 4.46 g/min feeding rate, 6.83 g/min steam flow rate), total energy requirements for a 1 ton/h capacities were about 600,000 kWh/kg and 3129 kWh/ton respectively (Eimontas et al. 2022; Nam et al. 2023).

An economic analysis performed on a pyrolysis plant generated products yields and costs of about 42.26% bio-oil (US$ 0.30/kg), 54.13% gas (US$ 0.2/kg), and 3.61% biochar (US$1.2/kg) (Badger et al. 2011; Eimontas et al. 2022; Kodera et al. 2021). The cost of syngas (the main product) from the gasification unit was comparatively higher (US$ 3.29/GJ) compared to the cost of bio-oil (main production in pyrolysis) from the pyrolysis unit (Pei et al. 2016; Kodera et al. 2021). Pei et al. (2016) reported that the gasification system's TEA revealed a high performance in the treatment of biomedical waste with revenue generation beginning at year 8 with a net present value of €4.59 million, 14-year discounted payback time and 12.51 internal rate of return (Yousef et al. 2023).

A 200 kg/h plastic waste pyrolysis plant with reactor operating at 650 °C and 1–3.1 kg/h feeding rate (yielding 54.7–69.1 wt% pyrolysis gas) was simulated to obtain economic indices for a TEA study which showed that it costs less (US$ 177,000/year) to adopt plastic waste pyrolysis gas as a natural gas replacement for a cogeneration system (total annual energy consumption of $1.87 \times 10^7$ MJ/year) with annual cost of plant, annual operation cost, and annual economic value of fuel gas of US$ 246,700, US$ 316,900, and US$ 291,700 respectively (Kodera et al. 2021).

Catalytic (Platinum Pt, and Ruthenium, Ru) hydrothermal gasification of carbohydrate-rich waste/wastewater for simultaneous hydrogen production and waste treatment in a fixed bed reactor was evaluated economically using the ASPEN Plus model (described in previous sections of this chapter) and results showed a minimum selling price of $H_2$ to be US$ 3.4/kg where feedstock price and amount contributed significantly to the cost (Paida et al. 2019). A dual catalyst reactor (e.g. Pt-Ru) is found to minimize reactor cost by about 40% because the desired results of high $H_2$ yields and high carbon gasification are combined with 2 catalysts (Paida et al. 2019). Using off-gases from gasification and pyrolysis units for electricity production for use in the plant and obtaining by-product credits can result in excess electricity which can be sold to the grid at about US$ 0.0775/kWh (Paida et al. 2019).

Waste cooking oil has been found to be a socio-economically sustainable feedstock for pyrolysis compared to conventional technologies (Mohanty et al. 2024; Tao et al. 2017). Depending on a specific region, low cost of feedstock of waste cooking oil (US$ 0.23/kg) contributed to a very low minimum selling price (US$0.91/L) of aviation fuel generated via pyrolysis of waste cooking oil (Mohanty et al. 2024). Comparing the cost of transportation fuel ($H_2$) generated via pyrolysis and gasification of corn stover, Anex et al. (2010) concluded that pyrolysis was a better technology with the least total capital investment and production cost (US$ 2.0/gallon) as opposed to gasification with high product cost of US$ 5.50/gallon. Depending on waste composition, co-pyrolysis can be economically advantageous to gasification because of the positive influence different waste materials can have on pyrolysis procedures and products without product cost escalation (Mohanty et al. 2024). The generation of woody biomass (e.g., pine, poplar waste) pyrolysis products (with upgrade/hydrodeoxygenation unit) is very costly although the technology produces high biofuel yields compared to herbaceous biomass (Meyer et al. 2020). However, co-pyrolysis of woody and herbaceous biomass is reported to be technically and economically feasible. Again, a stationary facility of co-pyrolysis of corn cob, sawdust, wheat straw, and pinewood mix in a 100 kg/h capacity fluidizes be reactor with pretreatment unit was compared to a mobile one via ASPEN Plus TEA model with results indicating better economic performance of the mobile pyrolysis unit because of its high turnover (Chen et al. 2018). Also, microwave-assisted vacuum co-pyrolysis of waste cooking oil and plastic waste mix generated a low oxygenated compounds with high amount of light hydrocarbons having very high oil yield of 84 wt% with high higher heating value (HHV) of 49 MJ/kg at a lower bio-oil cost of US$ 2.02/l compared to US$ 2.54/L when the process was carried out in a $N_2$ environment (Lam et al. 2019). The scale-up of the waste cooking oil-plastic waste mix co-pyrolysis is found to remarkably reduce the cost of bio-oil (US$ 2.02/L) by about 88% which could be competitive with fossil-based diesel (Lam et al. 2019). Similarly, the scale-up of a pyrolysis plant converting forest residue into pyrolysis oil was found to have a notable decrease in selling price of bio-oil by about 35% (van Schalkwyk et al. 2020).

Forest residues can be efficiently pyrolyzed with hydrothermal treatment to generate diesel/gasoline hydrocarbons (61% bio-oil yield, 24% char yield, and 15% gas yield) with minimum selling price of US$ 6.25/gallon considering a 30-year project duration, a US$ 427 million TCI and US$ 154 million annual operational cost at 2000 dry metric tons of waste per day facility (Carrasco et al. 2017). Biodiesel production from corncob pyrolysis in a fluidized bed reactor with hydrotreating unit at a capacity of 96.50 tons/h was simulated in PYROL HYSYS, and TEA performed on the technology showed a minimum selling price of the diesel at US$1.47/gallon which is very low compared to diesel obtained from forest residues hydrothermal pyrolysis at a higher plant capacity (Brigagão et al. 2019). Similarly, rice husk pyrolysis in a fluidized bed reactor at a smaller production capacity (4000 kg/h) yielded liquid fuel oil at a selling price of 1748 yuan/ton (i.e., US$ 258.704/ton[3] $\simeq$ US$ 0.259/kg) which is very low compared to the expected selling price of pyrolysis liquid fuel in China in 2017 (2100 yuan per ton $\simeq$ US$ 310.8/ton $\simeq$ US$ 0.311/kg) (Ji et al. 2017). Comparing these costs with larger size pyrolysis plant (1000 tons/day $\simeq$ 41,667 kg/h) using same feedstock (rice husks), it was deduced from a TEA study by Wang and Jan (2018) in agreement with results from Lam et al. (2019) that scale-up reduced the cost of bio-oil notably to about US$ 0.55/L ($\simeq$ US$ 0.57/kg[4]). A demonstration pyrolysis plant was assessed for its commercial economic viability in Thailand, and it was indicated that with upgrading process of the bio-oil and government subsidies, it would have a high potential to generate attractive economic returns (Jaroenkhasemmeesuk and Tippayawong 2015). A 3-in-1 novel design comprising waste tire pyrolysis, silicon production waste-heat recovery, and organic Rankine cycle was developed by Zhao et al. (2023), and TEA results showed a short payback time (which would be changing) of 3.79 years. Gasification of waste

[3] This is a conversion of 1748 Chinese Yuan (CNY) to US Dollar at a 2017 average exchange rate of US$ 0.1480 per 1 CNY.

[4] Assuming 1 Liter of bio-oil equals 0.951 kg.

tire for the production of liquefied synthetic natural gas (SNG) using rotary kiln gasifier was assessed for three different scenarios (without $CO_2$ capture and sequestration; with precombustion carbon capture sequestration and with pre- and post-combustion carbon capture sequestration) in two different locations (United States and Norway) using TEA, and results showed a lower minimum selling price of the low heating value SNG at US$16.7/GJ in United States and US$ 20.9/GJ in Norway via the system without carbon capture (Subramanian et al. 2020). They concluded that gasification system with pre-and post-combustion units located in Norway has the worst economic performance (minimum selling price of US$ 24.9/GJ) compared to the same scenario in the United States (minimum selling price of US$ 19.9/GJ), and these prices may be competitive under specific regulatory conditions (e.g., Canada requiring utilities companies to buy bio-SNG at prices up to $30/GJ), scale-up and levying of tipping fees (Subramanian et al. 2020).

Hybrid gasification waste-to-energy system (consisting of 245 kW photovoltaic, 175 kW wind, 250 kW biomass gasifier, and 46 batteries) utilizing 3500 tons of municipal solid waste (MSW) to generate 400 kW of renewable electricity and 500 kW of clean heat per day was found to be economically viable with electricity cost of 0.0941 CAD[5]/kWh ($\simeq$ US$ 0.0751 kWh) and operating cost of 44,731 CAD/year ($\simeq$ US$ 35,679.2/year) (Esfilar et al. 2021). TEA for a hybrid pyrolysis with anaerobic digestion of biomass residues-cattle manure mix at a capacity of 328,716 ton/year resulted in a 4.4-year payback period, 110,000 ton/year break-even point, 23% return on investment and net profit of €3.1 million/year (Aravani and Papadakis 2024).

With higher internal rate of return values compared to 10% assumed interest rate for a waste-to-energy plant in Greece (using tree pruning biomass), the facility was found to be economically beneficial with a payback period of 3.5–5 years and provision of employment opportunities (Sagani et al. 2019). An increase in the sales price of energy, reduction in investment cost, and reduction in biofuel price for this facility contributed significantly to the economic sustainability of the plant.

Waste-to-energy systems in developing countries are attracting interest due to their profitability, job creation abilities, and lower cost of electricity generated. The cost assessment performed on a MSW incineration systems located in Ghana indicated a cheaper average cost of electricity (US$ 0.039/kWh) for the integrated system with already existing closed engineered configuration (Ofori-Boateng et al. 2013).

To produce ethylene and jet fuel by thermal treatment of agricultural waste such as rice straw at a production capacity of 100 dry ton/h, about 27.8 MWh energy would be required to produce 1 ton of products at 43.7% energy efficiency at minimum selling prices of US$ 1.34/kg ethylene and US$1.52/L jet fuel (Michailos and Webb 2020). Sensitivity analysis for this system revealed capital cost and feedstock cost-intensive processes where a slight rise in the yield of the products would shift the economic performance to the positive direction.

[5] CAD (Canadian dollar) conversion to US dollar at an average exchange rate of US$ 0.797639 in 2021.

Factors that have significantly influenced the economic feasibility of pyrolysis waste and gasification technologies include (Aravani and Papadakis 2024; Zepeda et al. 2024; Rojas et al. 2024):

- Products selling price variation, which can change the economic dynamics of the plant. Zepeda et al. (2024) concluded that for a rice straw biochar-based slurry fuel plant, a 10% raise in minimum selling price of a 40 wt% yield biochar in water slurry and biochar in ethanol slurry led to about 98.8% probability of economic feasibility.
- Raw materials costs (including gasifying agents)
- Production capacity
- Incentive cost
- Total fixed capital
- Energy cost
- Capital recovery factor
- Internal Rate of Return (IRR)

## 6.4 Conclusion

The technoeconomic analysis results of thermochemical waste conversion technologies specifically pyrolysis, gasification, hydrothermal treatment, and combustion of palm were discussed and compared with fossil-based conversion technologies. Biofuels from thermochemical conversion technologies are found to be highly competitive with fossil diesel in terms of cost though they require high initial capital investment (TCI) for commercial scale facilities. Scaled up waste cooking oil-plastic waste mix co-pyrolysis, for instance, was found to drastically reduce the cost of bio-oil (US$ 2.02/L) by about 88% which could be competitive with fossil-based diesel. Sensitivity analysis of thermochemical waste conversion technologies revealed capital cost and feedstock cost-intensive processes where an increase in the yield of the products would shift the economic performance to the positive direction.

## References

Anex RP, Aden A, Kazi FK, Fortman J, Swanson RM, Wright MM, Satrio JA, Brown RC, Daugaard DE, Platon A, Kothandaraman G, Hsu DD, Dutta A (2010) Techno-economic comparison of biomass-to-transportation fuels via pyrolysis, gasification, and biochemical pathways. Fuel 89(1):S29–S35

Aravani VP, Papadakis VG (2024) Technoeconomic assessment for the viable exploitation of biomass residues by an innovative pyrolysis-anaerobic digestion processing plant. Environ Res 243:117835

Badger P, Badger S, Puettmann M, Steele P, Cooper J (2011) Techno-economic analysis: Preliminary assessment of pyrolysis oil production costs and material energy balance associated with a transportable fast pyrolysis system. BioResources 6:6048

Boie W (1953) Fuel Technology Calculations. Energ Tech 3:309–316
Brigagão GV, de Queiroz FAO, de Medeiros JL, Mikulcic H, Duic N (2019) A techno-economic analysis of thermochemical pathways for corncob-to-energy: fast pyrolysis to bio-oil, gasification to methanol and combustion to electricity. Fuel Process Technol 193:102–113
Carrasco JL, Gunukula S, Boateng AA, Mullen CA, DeSisto WJ, Wheeler MC (2017) Pyrolysis of forest residues: An approach to techno-economics for bio-fuel production. Fuel 193:477–484
Chen X, Zhang H, Xiao R (2018) Mobile autothermal pyrolysis system for local biomass conversion: Process simulation and techno-economic analysis. Energy Fuels 32:4178–4188
Danish Energy Agency (2023) Technology data—energy plants for electricity and district heating generation. Danish Energy Agency, København, Denmark. https://ens.dk/en/our-services/technology-catalogues. Accessed 26 Apr 2024
Do TX, Lim Y (2016) Techno-economic comparison of three energy conversion pathways from empty fruit bunches. Renew Energy 90:307–318
Economopoulos AP (2010) Technoeconomic aspects of alternative municipal solid wastes treatment methods. Waste Manage 30(4):707–715
Eimontas J, Stasiulaitiene I, Zakarauskas K, Striūgas N (2022) Pyrolysis of all layers of surgical mask waste as a mixture and its life-cycle assessment. Sustain Prod Consum 32:519–531
Esfilar R, Bagheri M, Golestani B (2021) Technoeconomic feasibility review of hybrid waste to energy system in the campus: A case study for the University of Victoria. Renewable and Sustainable Energy Rev 146:111190
Fadhilah NA, Islam MN, Rosli R (2023) Techno-economic analysis of sawdust and rice husk co-pyrolysis for bio-oil production. Bioresour Technol Rep 21:101233
Farzad S, Mandegari M, Görgens JF (2021) A novel approach for valorization of waste tires into chemical and fuel (limonene and diesel) through pyrolysis: Process development and techno economic analysis. Fuel Process Technol 224:107006
Gul E, Riva L, Nielsen HK, Yang H, Zhou H, Yang Q, Skreiberg Ø, Wang L, Barbanera M, Zampilli M et al (2021) Substitution of coke with pelletized biocarbon in the European and Chinese steel industries: An LCA analysis. Appl Energy 304:117644
Hogg D (2002) Costs for municipal waste management in the EU. https://ec.europa.eu/environment/pdf/waste/studies/eucostwaste.pdf. Accessed 12 July 2023
Hydrogen Council (2020) Path to hydrogen competitiveness—a cost perspective hydrogen council. Brussels, Belgium, pp 1–88
Jain S, Sharma MP (2010) Prospects of biodiesel from Jatropha in India: a review. Renew Sust Energ Rev 14:763–771
Jaroenkhasemmeesuk C, Tippayawong N (2015) Technical and economic analysis of a biomass pyrolysis plant. Energy Procedia 79:950–955
Ji LQ, Zhang C, Fang JQ (2017) Economic analysis of converting of waste agricultural biomass into liquid fuel: a case study on a biofuel plant in China. Renew Sustain Energy Rev 70:224–229
Kannah RY, Kavitha PS, Karthikeyan PO, Kumar G, Dai-Viet NV, Banu JR (2021) Techno-economic assessment of various hydrogen production methods—a review. Bioresour Technol 319:124175
Khan SR, Zeeshan M, Fatima S, Ciolkosz D, Dimitriou I, Jin H (2023) A comparative techno-economic analysis of combined oil and power production from pyrolysis and co-pyrolysis plants utilizing rice straw and scrap rubber tires. Fuel 348:128639
Kodera Y, Yamamoto T, Ishikawa E (2021) Energy- and economic-balance estimation of pyrolysis plant for fuel-gas production from plastic waste based on bench-scale plant operations. Fuel Commun 7(3):100016
Lam SS, Mahari Wan WA, Ok YS, Peng W, Chong CT, Ma NL, Chase HA, Liew Z, Yusup S, Kwon EE, Tsang DCW (2019) Microwave vacuum pyrolysis of waste plastic and used cooking oil for simultaneous waste reduction and sustainable energy conversion: Recovery of cleaner liquid fuel and techno-economic analysis. Renew Sustain Energy Rev 115:109359
Lee KT, Ofori-Boateng C (2013) Sustainability of biofuel production from oil palm biomass. Springer Science+Business Media, Singapore, pp 323

Lim Y, Lee H, Lee Y, Han C (2009) Design and economic analysis of the process for biodiesel fuel production from transesterificated rapeseed oil using supercritical methanol. Indust Eng Chem Res 48:5370–5537

Meyer PA, Snowden-Swan LJ, Jones SB, Rappé KG, Hartley DS (2020) The effect of feedstock composition on fast pyrolysis and upgrading to transportation fuels: techno-economic analysis and greenhouse gas life cycle analysis. Fuel 259:116218

Michailos S, Webb C (2020) Valorization of rice straw for ethylene and jet fuel production: a technoeconomic assessment. In: Kosseva MR, Webb C (eds) Food industry wastes, 2nd edn. Academic Press, pp 201–221

Michailos S, Parker D, Webb C (2017) A techno-economic comparison of Fischer-Tropsch and fast pyrolysis as ways of utilizing sugar cane bagasse in transportation fuels production. Chem Eng Res Design 118:206–214

Mohanty A, Ajmera S, Chinnam S, Kumar V, Mishra RK, Acharya B (2024) Pyrolysis of waste oils for biofuel production: An economic and life cycle assessment. Fuel Commun 18:100108

Nam JY, Lee TR, Tokmurzin D, Park SJ, Ra HW, Yoon SJ, Mun TY, Yoon SM, Moon JH, Lee JG et al (2023) Hydrogen-rich gas production from disposable COVID-19 mask by steam gasification. Fuel Lond 331:125720

OECD (Organization for Economic Co-operation and Development) (2001) Analytic Report on Sustainable Development SG/SD (2001)1–14, OECD, Paris. https://www.oecd.org/sd-roundtable/papersandpublications/39370725.pdf. Accessed 6 Mar 2023

Ofori-Boateng C, Lee KT, Mensah M (2013) The prospects of electricity generation from municipal solid waste (MSW) in Ghana: A better waste management option. Fuel Process Technol 110:94–102

Ouda O, Raza S, Nizami A, Rehan M, Al-Waked R, Korres N (2016) Waste to energy potential: a case study of Saudi Arabia. Renew Sustain Energy Rev 61:328–340

Paida VR, Kersten SRA, van der Ham AGJ, Brilman DWF (2019) A two-step approach to the hydrothermal gasification of carbohydrate-rich wastes: Process design and economic evaluation. Int J Hydrog Energy 44(47):25524–25541

Pei P, Korom SF, Ling K, Nasah J (2016) Cost comparison of syngas production from natural gas conversion and underground coal gasification. Mitig Adapt Strateg Glob Chang 21:629–643

Peters MS, Timmerhaus KD (1981) Plant design and economics for chemical engineers, 3rd edn. McGraw-Hill Book Company, pp 143

Rojas J, Zhai S, Eddie S, Vasudev H, Marin-Quiros S, et al (2024) Technoeconomics and carbon footprint of hydrogen production. Int J Hydrog Energy 49(Part D):59–74

Sagani A, Hagidimitriou M, Dedoussis V (2019) Perennial tree pruning biomass waste exploitation for electricity generation: the perspective of Greece. Sustain Energy Technol Assessments 31:77–85

Seider WD, Seader JD, Lewin DR, Widagdo S (2008) Product and process design principles: synthesis, analysis and evaluation, 3rd edn. Wiley

Sharma A, Jakhete A, Sharma A, Joshi JB, Pareek V (2019) Lowering greenhouse gas (GHG) emissions: techno-economic analysis of biomass conversion to biofuels and value-added chemicals. Greenhouse Gases: Sci Technol 9:454–473

Sinnot RK (1986) An introduction to chemical engineering design. Chem Eng 33:116–117

Subramanian ASR, Gundersen T, Adams TA (2020) Technoeconomic analysis of a waste tire to liquefied synthetic natural gas (SNG) energy system. Energy 205:117830

Tao L, Markham JN, Haq Z, Biddy MJ (2017) Techno-economic analysis for upgrading the biomass-derived ethanol-to-jet blendstocks. Green Chem 19:1082–1101

Tomomatsu Y, Swallow B (2007) Jatropha curcas Biodiesel Production in Kenya: Economics and Potential Value Chain Development for Smallholder Farmers, Working Paper 54, World Agro forestry Centre, pp 33

Turton R, Bailie C, Whiting B, Shaeiwitz A (2009) Analysis, Synthesis and design of chemical processes. Third edition, Pearson Education Inc., Boston, MA USA

van Schalkwyk DL, Mandegari M, Farzad S, Görgens JF (2020) Techno-economic and environmental analysis of bio-oil production from forest residues via non-catalytic and catalytic pyrolysis processes. Energy Convers Manag 213:112815
Wang WC, Jan JJ (2018) From laboratory to pilot: Design concept and techno-economic analyses of the fluidized bed fast pyrolysis of biomass. Energy 155:139–151
Wang Y, Zhang S (2017) Economic assessment of selected hydrogen production methods: A review. Energ Source Part B Econ Planning Policy 12:1022–1029
Wang Y, Li G, Liu Z, Cui P, Zhu Z, Yang S (2019) Techno-economic analysis of biomass-to-hydrogen process in comparison with coal-to-hydrogen process. Energy 185:1063–1075
You YD, Shie JL, Chang CY, Huang SH, Pai CY, Yu YH et al (2008) Economic cost analysis of biodiesel production: case in soybean oil. Energy Fuel 22:182–189
Yousef S, Lekavičius V, Striūgas N (2023) Techno-economic analysis of thermochemical conversion of waste masks generated in the EU during COVID-19 pandemic into energy products. Energies 16:3948
Zepeda LC, Al-Waili I, Griffin G, Shah K, Bhargava S, Parthasarathy R (2024) Techno-economic assessment (TEA) and sensitivity analysis for a biochar-based slurry fuel plant. Energy Convers Manag 306:118297
Zhao X, Chen H, Li S, Li W, Pan P, Liu T, Wu L, Xu G (2023) Thermodynamic and economic analysis of a novel design combining waste tire pyrolysis with silicon production waste heat recovery and organic Rankine cycle. Energy 283:128500

# Index




C. Ofori-Boateng, *Sustainability of Thermochemical Waste Conversion Technologies*, https://doi.org/10.1007/978-3-031-64342-2

The manufacturer's authorised representative in the EU is Springer Nature Customer Service Centre GmbH, Europaplatz 3, 69115 Heidelberg, Germany. If you have any concerns regarding our products, please contact ProductSafety@springernature.com

Printed and bound by CPI Group (UK) Ltd, Croydon, CR0 4YY
01/10/2026
02225961-0004